THE SYSTEMIC PRACTICE
OF MISINTERPRETATION
OF SCIENTIFIC DATA

THE SYSTEMIC PRACTICE OF MISINTERPRETATION OF SCIENTIFIC DATA

THE CASE OF PERSISTERS, SMALL COLONY VARIANTS, VIABLE BUT NON-CULTURABLE BACTERIA, AND SENESCENT BACTERIA IN MICROBIOLOGY

JAISON JACOB

Universal-Publishers
Boca Raton

The Systemic Practice of Misinterpretation of Scientific Data:
The Case of Persisters, Small Colony Variants, Viable but Non-Culturable
Bacteria, and Senescent Bacteria in Microbiology

Universal-Publishers
Boca Raton, Florida • USA
2010

ISBN-10: 1-59942-820-2
ISBN-13: 978-1-59942-820-8

www.universal-publishers.com

TABLE OF CONTENTS

Abbreviations VII

Introduction IX

CHAPTER I

Fundamental Concepts in Pharmacokinetics
and Pharmacodynamics 13

CHAPTER II

Persisters, Phenotypic Shift and Chronic Infections 37

Persisters: A Review of Literature 39

Why is Phenotypic Shift of Persisters an In Vitro Illusion? 51

CHAPTER III

Small Colony Variants and Chronic Infections 73

Small Colony Variants: Current Knowledge 75

Small Colony Variants: Are They Responsible for Chronic Infections? 84

CHAPTER IV

Resuscitation of Viable but Non-Culturable Bacteria,
Outbreak and Global Spread of Cholera 99

Viable but Non-Culturable Bacteria: A Review 101

VBNC: A Tale of Two Illusions 125

CHAPTER V

Senescent Bacteria and Models of Aging 155

Bacterial Senescence: A Review 157

Why the Current Model of E. Coli Aging is Incomplete 170

CHAPTER VI

Integrating Persisters, SCV, VBNC, and Senescent Bacteria
with PK/PD Parameters 177

*Persisters Show Heritable Phenotype and Generate Bacterial Heterogeneity
and Noise in Protein Expression* 179

A New Model of Bacterial Aging Incorporating SCVS, VBNC, and Persisters 202

Senescent Bacteria as Potential Live Vaccines 212

CHAPTER VII

Conclusions and Predictions 221

CHAPTER VIII

Creating Illusions 227

Index 237

ABBREVIATIONS

ASW	Artificial sea water
AUC	Area under the concentration-time curve
CF	Cystic fibrosis
Cmax	Maximum concentration
ESBL	Extended spectrum beta lactamase
GFP	Green fluorescent protein
HNA	High nucleic acid
LB	Luria Bertani
LNA	Low nucleic acid
MBC	Minimum bactericidal concentration
MIC	Minimum inhibitory concentration
MPC	Mutant prevention concentration
MSSA	Methicillin sensitive Staphylococcus aureus
MSW	Mutant selection window
OD	Optical density
PAE	Post antibiotic effect
PCD	Programmed cell death
PD	Pharmacodynamics
PK	Pharmacokinetics
SCV	Small colony variants
SOD	Superoxide dismutase
TA	Toxin-antitoxin
TD-SCV	Thymidine-dependent small colony variants
t>MIC	Percentage of time above MIC
T_{MSW}	Time inside mutant selection window
VBNC	Viable but non-culturable bacteria

INTRODUCTION

Scientific literature is a record of pursuing the truth (Sox and Rennie 2006). In business or politics, some level of dishonesty or corruption is expected or can be overlooked as it is generally accepted as an unavoidable part of the profession. On the other hand, scientific research is done to find the truth and hence dishonesty even at small levels contradicts its fundamental aim (Beisiege 2010). It is a general belief that the scientific community is relatively free of corruption and adopts good scientific practices. However, science in recent times has become a big business and has been influenced by financial interests and pressure groups (Resnik 2007). Billions of dollars are spent each year for research and development by governments and private industries. Being the owners of companies and intellectual property rights, many scientists now have a financial stake in research (Resnik 2007). Moreover, universities have commercial interests as they can patent their employees' research and collect royalties.

Research misconduct is on the rise and, in a recent survey, it was reported that more than a third of US scientists have engaged in scientific misconduct (Wadman 2005). Research misconduct generally includes data falsification, fabrication and plagiarism (FFP) (Martyn 2003). Even though explicit FFP are rare, misconduct other than FFP can be more common (Martinson *et al.* 2005). It is also argued that research that is not needed or that constitutes poor value for the money should be considered as misconduct (Smith 2008) since most of the money spent by researchers is produced by the taxpayer.

Many scientists obtain funding through government grants and contracts, which are also an important source of revenue for the universities (Resnik 2007). The indirect cost included in the grants (which is approximately 30% of the direct cost) is intended to cover the administrative costs associated with the research activities (Resnik 2007). Since the left over money can be used for other purposes, universities benefit from obtaining more research grants. It is no wonder that the scientists who bring in the most grant money have a greater influence in their department and in the university. Thus, obtaining funding from the government or from private sources is very important for scientists to remain in their research field. The pressure to obtain funding may encourage at least some of them to falsify or fabricate data during grant applications. The same pressure may also be at work during the submission of articles to scientific journals, since the number of publications is generally considered to be a measure of research productivity. Even though the individual researcher or the university benefits from the situation, the scientific community and the taxpayer become the ultimate losers.

In this book, I have focused on how researchers misinterpret their data and create illusions with the intention of developing an attractive hypothesis. Through illusions, the hypothesis is made attractive and subsequently developed into an acknowledged scientific truth following their publication in peer-reviewed journals. Later, it becomes difficult to challenge those findings and thus they remain in the scientific literature, uncorrected, for long periods of time, resulting in significant loss of money, time and resources. Even though the hypothesis may appear to be true, it is possible to identify the flaws by scrutinizing the subject very closely, a task expected to be accomplished by journal reviewers. However, many times, this function is not carried out precisely, resulting in the publication of distorted scientific views.

The main focus of my book is to show how researchers have exaggerated the ability of bacteria to cause chronic infections and disease outbreaks. Even though bacteria do possess the ability to adjust to unfavorable conditions, I argue that researchers have exaggerated it beyond the point of truth, probably to get more attention from the scientific community.

I start with some of the fundamental concepts in pharmacokinetics and the pharmacodynamics of antibiotics (topics necessary for studying bacterial killing kinetics but is almost completely ignored by molecular microbiologists). Then, I discuss four related but independent topics: persisters, small colony variants, viable but nonculturable bacteria and senescent bacteria. Each topic is divided into two sections; first, a review of the literature; and second, questioning the validity of the current hypothesis and findings. In the subsequent chapters, a simpler hypothesis is offered based on my research findings, after integration of all four topics. Finally, the impact of creating illusions in research is also discussed.

References

Beisiege, U. (2010). Research integrity and publication ethics. *Atherosclerosis* doi:10.1016/j.atherosclerosis.2010.01.050

Martinson, B. C., Anderson, M. S., and de Vries, R. (2005). Scientists behaving badly. *Nature* 435(7043), 737-738.

Martyn, C. (2003). Fabrication, falsification and plagiarism. *Qjm-an International Journal of Medicine* 96(4), 243-244.

Resnik, D. B. (2007). "The price of truth: How money affects the norms of science." Oxford University Press, Oxford, U. K.

Smith, R. (2008). Most cases of research misconduct go undetected, conference told. *BMJ* 336 (7650), 913.

Sox, H. C., and Rennie, D. (2006). Research misconduct, retraction, and cleansing the medical literature: Lessons from the Poehlman case. *Annals of Internal Medicine* 144(8), 609-613.

Wadman, M. (2005). One in three scientists confesses to having sinned. *Nature* 435(7043), 718-719.

CHAPTER I

FUNDAMENTAL CONCEPTS IN PHARMACOKINETICS AND PHARMACODYNAMICS

P harmacokinetics (PK) uses mathematical models to study the movement of drugs through the body. It predicts the time course of drug concentrations and its metabolites within the body (Craig 1998; Burgess 1999). Pharmacokinetics is concerned with the absorption, distribution, biotransformation and elimination of the drug and thus determines the onset, duration and intensity of the drug action. In simple terms, it studies the fate of the drug inside the body. Some of the important PK parameters are the half-life of the drug (t1/2), the area under the concentration-time curve (AUC), peak concentration (Cmax), bioavailability and volume of distribution. Pharmacodynamics (PD), on the other hand, is related to the drug action and its effects i.e. how the drug binds the receptors, the effect of the drug after receptor binding, and its effect on the organism (Woodnutt 2000; Cazzola and Matera 1998). The most important PD parameters are the minimum inhibitory concentration (MIC) and the minimum bactericidal concentration (MBC). The combined PK/PD studies the dose-effect relationship of the drug (Craig 2001; Cazzola and Matera 1998). Important PK/PD parameters that determine the efficacy of antibiotics include the percentage of time above the MIC (t>MIC), Cmax/MIC and AUC/MIC. The knowledge on PK/PD allows the selection of optimal dosage regimen and helps to predict the clinical outcome and prevent the emergence of resistant organisms (Craig 2001; Cazzola and Matera 1998).

For an antibiotic to eradicate organisms, it should bind to the target sites in bacteria (cell wall components, protein synthesis machinery, enzymes etc.) and occupy sufficient binding sites (which is dependent on the concentration of antibiotics) for an adequate period of time. Thus, both the concentration and the time determine the efficacy of bacterial eradication by the antibiotics (Cazzola and Matera 1998; Craig 1998). Even though MIC and MBC (PD parameters) give a general idea about the antibiotic's efficacy, they can not give any indication about the time course of antibacterial activity nor can they account for the fluctuations in the concentration of antibiotics within the body (Mueller et al. 2004; Craig 1998; Jacobs 2003). Moreover, the antimicrobial activity tested in liquid media in the absence of host factors can be considerably different from the clinical outcome (Turnidge 1998). However, PK/PD parameters can provide a much better prediction for *in vivo* antibacterial activity (Cazzola and Matera 1998; Craig 1998; Jacobs 2003).

Antibiotics differ in the PK/PD parameter that best correlates with the therapeutic efficacy (Craig 1998; Woodnutt 2000; Mueller *et al.* 2004). For β-lactam antibiotics, most of the macrolides, and clindamycin, the single most important parameter that determines efficacy is the percentage of time that the drug concentration remains above MIC (t>MIC) (Cazzola and Matera 1998; Craig 1998; Jacobs 2003; Ambrose *et al.* 2007). This class of antibiotics is referred to as time-dependent antibiotics because the duration of exposure to antibiotics determines the extent of antibacterial activity. Here, achieving very high concentration of antibiotics is not required because concentrations above 3-4 times the MIC may not significantly enhance bacterial killing (Turnidge 1998; Cazzola and Matera 1998; Craig 1998). However, maintaining the antibiotic concentration above the MIC for sufficiently long periods of time is critical for therapeutic efficacy (Craig 1998). Antibiotics need not be maintained above MIC for the whole dosing interval, but t>MIC higher than 40-50% can result in a high bacteriological cure rate (Craig 2001). However, t>MIC may vary depending on the organism (Craig 1995). In animal infection studies with cephalosporins, maximal efficacy was noticed when t>MIC was 60-70% of the dosing interval for *Enterobacteriaceae* and *Streptococci* and 40%-50% for *Staphylococcus aureus* (Craig 1995). Pea and Viale (2006) suggest that, because of the low postantibiotic effect exhibited by this group, the drug concentration should not be allowed to reach trough levels below the MIC. In immunocompromised patients, t>MIC may have to be increased to 100% (Mouton and Vinks 1996). For this class of antibiotics, continuous infusion or small dosing intervals of antibiotics may help to maintain the plasma antibiotic concentration above the MIC for sufficiently long periods of time (Pea and Viale 2006; Navas *et al.* 2006; Mariat *et al.* 2006). When the half-life of the drug is shorter, more frequent dose fractioning must be followed (Pea and Viale 2006). Continuous infusion has the advantage of attaining more favorable PK/PD parameters than intermittent administration (Van Zanten *et al.* 2007; Navas *et al.* 2006; Mariat *et al.* 2006). Van Zanten *et al.* (2007) compared the efficacy of continuous and intermittent administration of cefotaxime in patients with chronic obstructive pulmonary disease and respiratory tract infections. Even though the clinical cure rates were comparable, it was noticed that the continuous administration had the benefit of attaining more favorable PK/PD parameters that can theoretically prevent resistant organisms. Antibiotic concentrations exceeded 5xMIC at all the times during the continuous administration. Thus, at least against infections that are difficult to treat due to the problem of emergence of resistance, a continuous infusion may be more effective than intermittent administration (Mouton and Vinks 1996; Van Zanten *et al.* 2007; Navas *et al.* 2006; Mariat *et al.* 2006). Antibiotics such as tetracyclines and vancomycin also exhibit time-dependent killing (Cazzola and Matera 1998; Jacobs 2003; Craig 1998). However, they show prolonged and persistent effects. In this case also, the total exposure time is the important parameter of efficacy, but the concentration need not be above the MIC throughout the dosing interval to have the maximal bactericidal efficacy (Cazzola and Matera 1998; Jacobs 2003). In this case, AUC/MIC is the major PK/PD parameter that correlates with therapeutic efficacy (Craig 1998).

For the third class of antibiotics like aminoglycosides and fluoroquinolones, therapeutic efficacy is correlated with AUC/MIC and Cmax/MIC (Cazzola and Matera 1998; Burgess 1999; Craig 1998; Ambrose *et al.* 2007). This group of antibiotics is known as concentration-dependent antibiotics. Cmax/MIC is a better predictor for aminoglycoside efficacy (Kashuba *et al.* 1998), whereas AUC/MIC predicts better for fluoroquinolones and glycopeptides (Rybak 2006). The aim of the therapy should be to maximize the Cmax/MIC or AUC/MIC value (Cazzola and Matera 1998; Burgess 1999; Craig 1998; Kashuba *et al.* 1998). Cmax/MIC can be increased by maximizing the initial dose of the antibiotic (Burgess 1999). Thus, with aminoglycosides, the critical point to consider is that the initial dose should be given as high as possible, but below the level of toxicity (Burgess 1999). A Cmax/MIC ratio of 10-12 predicts successful clinical outcomes for aminoglycosides and fluoroquinolones and hence this ratio should be achieved at target sites for a favorable outcome (Burgess 1999; Kashuba *et al.* 1998; Preston *et al.* 1998; Ambrose *et al.* 2007). A high dose of antibiotic taken once daily can maximize the Cmax/MIC value (Burgess 1999; Pea and Viale 2006). Toxicity of these drugs, however, may limit the use of very high concentrations. In treatment of infections caused by bacteria exhibiting high MIC against aminoglycosides, it may not be possible to attain a Cmax/MIC value of 10-12 (Burgess 1999). In such cases, it may result in the emergence of a resistant population (Burgess 1999; Pea and Viale 2006). On the other hand, AUC/MIC can be maximized by either increasing the total daily dose of the antibiotic or by using an antibiotic with a low clearance rate such as azithromycin (Van Bambeke and Tulkens 2001). For fluoroquinolones, an

AUC/MIC ratio of 125 is associated with good clinical outcome against gram-negative organisms (Craig 1998; Schentag 2000) whereas a ratio of 30-40 may be sufficient against gram-positive organisms (Ambrose *et al.* 2001). An advantage with the concentration-dependent antibiotics is that they exhibit long post-antibiotic effects; hence sub-MIC levels can be allowed towards the end of the dosing intervals (Pea and Viale 2006).

Mutant prevention concentration and mutant selection window
Mutant prevention concentration (MPC), a relatively new concept in pharmacodynamics, is the lowest concentration of an antibiotic that can prevent the growth of a mutant population in a bacterial culture (Dong *et al.* 1999; Zhao and Drlica 2002; Drlica and Zhao 2007; Drlica 2003). When a bacterial culture is treated with increasing concentrations of fluoroquinolone, a sharp drop in the viable count of bacteria can be noticed initially, followed by a distinct plateau and subsequently by a second sharp decline in the viable count. The first sharp drop is due to the growth inhibition of wild type organisms, whereas the plateau is due to the presence of mutant subpopulations that are not inhibited by the antibiotic at MIC. However, as the concentration of antibiotic increases, the mutant subpopulations will also be inhibited, resulting in the second sharp drop. The concentration that prevents the growth of the first-step mutant subpopulation is the MPC (Dong *et al.* 1999; Zhao and Drlica 2002; Drlica and Zhao 2007; Drlica 2003). Organisms need to have two or more independent mutations to grow at concentrations above the MPC, the frequency of which is very low (Drlica and Zhao 2007). Any concentration between the MIC and MPC allows the selective growth of mutants, and this range of concentration between the MIC and MPC is the mutant selection window (MSW) (Drlica and Zhao 2007). MPC is determined by plating approximately 10^{10} bacterial cells on agar plates containing increasing concentrations of antibiotics (Zhao and Drlica 2002; Drlica and Zhao 2007). The concentration of the antibiotic that does not support the growth of any bacteria is taken as the MPC. Even though MIC and MPC are the concentrations that are responsible for the two sharp declines, there is no correlation between these two values (Lu *et al.* 2003; Hansen *et al.* 2006). For example, MPC/MIC varied from 12 to 130 when 5 different antibiotics were used against *Mycobacterium smegmatis* and from 3.2 to 64 against *S. aureus* (Lu *et al.* 2003). Similarly, the distinct plateau noticeable between the MIC and MPC may not be the same

for all antibiotic-pathogen combinations (Lu *et al.* 2003). For example, when penicillin was used against *M. smegmatis*, only an inflection point was detectable (Lu *et al.* 2003). The same was true when moxifloxacin, tetracycline, penicillin and chloramphenicol were used against *S. aureus* (Lu *et al.* 2003).

Optimizing the dose with relation to MPC as the pharmacodynamic parameter may not only help to achieve a therapeutic effect, but also prevent the emergence of a mutant population thus preventing drug resistance (Zhao and Drlica 2008). Maintaining the concentration of fluoroquinolones above the MPC can prevent the emergence of mutant strains, whereas the mutant population will be selected when the concentration of the antibiotic falls inside the MSW (Almeida *et al.* 2007; Ferran *et al.* 2009; Firsov *et al.* 2008; Croisier *et al.* 2004; Cui *et al.* 2006a). Selection of resistant bacteria occurred in a rabbit lung infection model, when the time inside MSW (T_{MSW}) was more than 45% of the total duration of treatment with gatifloxacin (Croisier *et al.* 2004). However, T_{MSW} as a whole may not be a good predictor for the selection of fluoroquinolone-resistant strains because it does not take into account the position of the antibiotic concentration within the MSW (Firsov *et al.* 2008; Cui *et al.* 2006a). The enrichment of mutants occurs frequently when the antibiotic concentration is close to the lower boundary of the MSW rather than to the upper boundary (Firsov *et al.* 2008; Cui *et al.* 2006a).

The PK/PD indices that could predict the prevention of resistant mutants have been studied both *in vitro* and *in vivo* (Croisier *et al.* 2004; Ferran *et al.* 2009; Firsov *et al.* 2008; Homma *et al.* 2007; Zhao and Drlica 2008; Olofsson *et al.* 2006; Olofsson *et al.* 2007). Using a mouse thigh bacterial infection model, the influence of marbofloxacin exposure on the selection of resistant *E. coli* was studied (Ferran *et al.* 2009). Among the different PK/PD parameters (AUC/MIC, Cmax/MIC and T_{MSW}), only T_{MSW} was found to be a good predictor of the prevention of resistance (Ferran *et al.* 2009). When T_{MSW} was higher than 34%, selection of the resistant population occurred, whereas the growth of resistant mutants was prevented when T_{MSW} was less than 30% of the treatment course (Ferran *et al.* 2009). In a related experiment, where the *in vivo* efficacy of gatifloxacin against *S. pneumoniae* in an experimental model of pneumonia was studied, it was found that the risk of mutation was very high when T_{MSW} was above 45% of the course of treatment (Croisier *et al.* 2004). However, a clear relationship between the emergence of a resistant population and T_{MSW} may not be noticeable in many cases

(Campion *et al.* 2004; Homma *et al.* 2007). Many researchers have pointed out that AUC/MPC and Cmax/MPC are better predictors than T_{MSW} in the prevention of resistant populations (Firsov *et al.* 2008; Homma *et al.* 2007; Zhao and Drlica 2008; Olofsson *et al.* 2006; Olofsson *et al.* 2007). When *S. pneumoniae* was exposed to different concentrations of moxifloxacin and levofloxacin, complete eradication of the organism occurred when AUC/MPC was above 13.41 and Cmax/MPC above 1.20 (Homma *et al.* 2007). When these values were below 0.84 and 0.08 respectively, susceptibility to the antibiotics was found to be decreased. Similarly, an AUC/MPC of 35 was sufficient to prevent the emergence of single-step mutants of *E. coli* (Olofsson *et al.* 2007). In terms of MIC, when AUC/MIC and Cmax/MIC were above 584 h and 47.2 respectively, emergence of *S. aureus* resistant population against ciprofloxacin was avoided, but not when these values were below 159 h and 13.4, respectively (Campion *et al.* 2004).

PK/PD indices that best correlate for β-lactam antibiotics for the prevention of resistance are not clear. In one experiment, it was shown that an AUC/MIC of 1000 can prevent the emergence of *E. cloacae* mutants against ceftizoxime (Stearne *et al.* 2007). However, in another experiment, no significant differences were noticed between MIC, MBC and MPC values when imipenem, meropenem, ceftriaxone, and ertapenem were used against three strains of *S. pneumoniae* (Hovde *et al.* 2003). The authors concluded that MPC may not be applicable to β-lactams that do not utilize a dual targeting system or to bacteria that utilize multiple resistance mechanisms. It should be noted that, even though MPC has been reported for antibiotics such as β-lactams, aminoglycosides, daptomycin and linezolid, the majority of data published on MPC are reported for fluoroquinolones. Smith *et al.* (2003) argued that, since chromosomal point mutations are not the primary resistance mechanisms for antibiotics other than fluoroquinolones, MPC may not be predictive for their activity. The authors suggest that MPC will be useful only when the mechanism of resistance *in vivo* corresponds to that evaluated in *in vitro* studies. However, other researchers argue that MPC may be applicable to antibiotics, including β-lactams and aminoglycosides, since the mutational resistance may not be limited to fluoroquinolones alone (Livermore 2003; Zhao 2003).

In vitro and *in vivo* results indicate that MSW may be important clinically since the traditional PD indices target the elimination of the susceptible population only, whereas they ignore the mutant populations that are selectively enriched and amplified (Epstein *et al.* 2004). Theoretically, PK/PD based on MPC may be helpful in the prevention of resistant populations. Similarly, MSW may be useful in the design and screening of new antibiotics with a narrow selection window so that the drug concentration remains inside MSW only for short periods of time (Epstein *et al.* 2004). However, its clinical usefulness is still doubtful due to a number of limitations in using MPC for direct clinical applications (Epstein *et al.* 2004; Zhao 2003). The presence of phenotypically tolerant population, difficulty in completely reconstituting a clinical setting for measurement of MPC, inability to kill resistant organisms and the lack of animal or human trials are some of the limitations in using MPC (Epstein *et al.* 2004; Zhao 2003). Moreover, doses required to prevent the growth of resistant populations can be much higher and thus carry the risk of adverse toxic effects (Epstein *et al.* 2004).

Inoculum effect
The inoculum effect is the increase in the MIC of an antibiotic when a higher inoculum size is used (Brook 1989; Craig *et al.* 2004). It was first reported in penicillinase-producing *Staphylococcus aureus* in which the inoculum effect was mainly due to the production of penicillinase enzyme that destroys the antibiotic (Benner *et al.* 1965; Sabath *et al.* 1975; Brook 1989). However, inoculum effect has been reported with other bacteria also (Balko *et al.*1999; Thomson and Moland 2001; Brook 1989; Soriano *et al.* 1996a; Kang *et al.* 2004). Traditionally, small inoculum sizes are used in determining the MIC or MBC values. However, during infections, the number of bacteria per ml of biological fluids can be much higher. Hence, some researchers argue that the MIC determined by the standard dilution method may not be appropriate *in vivo* (Soriano *et al.* 1990; Soriano and Ponte 2009).

The inoculum effect can be clearly demonstrated with β-lactam antibiotics, even though other antibiotics also show this effect (Brook 1989; Konig *et al.* 1998; Szabo *et al.* 2001; Morrissey and Smith 1994; Morrissey and George 1999). Even among the β-lactam groups, there are considerable differences between the antibiotics exhibiting the inoculum effect (Eng *et al.* 1985; Soriano *et al.* 1996b; Tam *et al.* 2009). It can be consistently detected with cefepime, cefotaxime, and ceftriaxone against extended spectrum β-lactamase (ESBL)-producing gram-negative pathogens, but only sometimes detected with piperacillin-tazobactam and infrequently detected with meropenem and cefoteten (Thomson and Moland 2001). Similarly,

Burgess and Hall (2004) noticed that piperacillin-tazobactam and cefepime were bactericidal against non-ESBL isolates of *Klebsiella pneumonia* at both standard and high inoculum sizes, but exhibited a significant inoculum effect against ESBL isolates, whereas meropenem and imipenem were bactericidal against both ESBL and non-ESBL isolates at any inoculum size. Similarly, the treatment failure in patients with aortic valve endocarditis caused by methicillin-susceptible *Staphylococcus aureus* was attributed to the inactivation of cefazolin by *S. aureus* type A β-lactamase (Nannini *et al.* 2003).

The inoculum effect can result from the ability of the bacteria to produce enzymes that hydrolyze the antibiotics (Craig *et al.* 2004). At high inoculum sizes, the initial bacterial killing may release more β-lactamases into the medium, hydrolyzing the antibiotics, resulting in more survivors and thus a higher MIC value (Craig *et al.* 2004). However, the inoculum effect can also result from reasons other than the ability of bacteria to produce enzymes that destroy the antibiotics (Stevens *et al.* 1993; Morrissey and Smith 1994; Udekwu *et al.* 2009). Stevens *et al.* (1993) hypothesized that during the stationary phase of bacterial culture growth, reduced expression of some penicillin-binding proteins (PBPs) can result in lower binding of antibiotics that may account for the inoculum effect. (However, this may be more related to the bacterial growth phase rather than to the inoculum effect itself. Whereas stationary phase bacteria were used in the above case, high cell density of exponential phase bacteria is generally used in experiments demonstrating the inoculum effect). The inoculum effect has also been attributed to reduced oxygen tension at high cell densities (Morrissey and Smith 1994). Moreover, the inoculum effect can also be noticed with non-ESBL pathogens and with organisms other than bacteria. A modest or substantial inoculum effect was noticed with six antibiotics (oxacillin, gentamicin, vancomycin, daptomycin, linezolid and ciprofloxacin) against β-lactamase-negative, methicillin-sensitive *S. aureus* (Udekwu *et al.* 2009). The inoculum effect for daptomycin and vancomycin was due to the loss of biological activity of the antibiotics at high cell densities, whereas for the other four antibiotics, it was due to a decrease in the antibiotic concentration per cell. Similarly, the inoculum effect is not restricted to bacteria alone (Steels *et al.* 2000; Gluzman *et al.* 1987; Takemura *et al.* 1991a; Takemura *et al.* 1991b; Kobayashi *et al.* 1992). *Zygosaccharomyces bailii* exhibited an inoculum effect against sorbic acid which was not due to metabolizing enzymes or adsorption of sorbic acid nor due to the lack of cell-to-cell signals in the medium (Steels *et al.* 2000). In addition, the inoculum effect of *Plasmodium falciparum* against chloroquine was not due to inactivating enzymes, but rather to reduced drug accumulation per parasite (Gluzman *et al.* 1987). Similarly, cytotoxicity of many anticancer drugs decreases with tumor cell densities (Takemura *et al.* 1991a; Takemura *et al.* 1991b; Kobayashi *et al.* 1992). The reduction in the cytotoxicity of both vincristine and doxorubicin was not due to the acidification of the medium at high cell density but was due to the lower number of drug molecules binding to the cellular target sites (Kobayashi *et al.* 1992). Takemura *et al.* (1991a, b) found a significant inoculum effect with doxorubicin but not with cisplatin. At high cell densities, there was reduction in doxorubicin accumulation in cells whereas cellular accumulation of cisplatin was almost the same at any density (Takemura *et al.* 1991a, b). Thus, reduction in drug accumulation by cells and loss of biological activity of antibiotics can be additional factors for an *in vitro* inoculum effect.

The inoculum effect exhibited by antibiotics like vancomycin can be indeed due to the loss of its biological activity (Yanagisawa *et al.* 2009; Sieradzki and Tomasz 2006; Cui *et al.* 2006b). When β-lactam-induced VAN-resistant and methicillin-resistant *S. aureus* culture was grown in the presence of vancomycin at 4μg/ml, the free vancomycin in the culture medium decreased to 2.3μg/ml in the first 8 h (Yanagisawa *et al.* 2009). However, cell growth was not detectable during this period indicating that the drug concentration was bacteriostatic. However, by 24 h, the vancomycin concentration was further reduced to approximately 1.5 μg/ml resulting in the re-growth of the cells. The gradual removal of vancomycin from the medium can be due to the trapping of the antibiotic molecules in the cell wall (Sieradzki and Tomasz 2006; Cui *et al.* 2006b) and/or due to the binding of antibiotics to the D-alanyl-D-alanine residues located in the cell wall which block murein hydrolases from attaching to its substrate, thus inhibiting cell wall lysis (Sieradzki and Tomasz 2006).

Antibiotics other than β-lactam types show modest to no inoculum effect, and the published findings show some inconsistencies in this regard. In general, antibiotics like aminoglycosides (Brook 1989; Konig *et al.* 1998; Szabo *et al.* 2001), fluoroquinolones (Konig *et al.* 1998; Firsov *et al.* 1999), carbapenems (Brook 1989; Konig *et al.* 1998; Kang *et al.* 2004) and linezolid (LaPlante and Rybak 2006; Udekwu *et al.* 2009) show minimal inoculum effects. LaPlante and Rybak (2006) studied the impact of high inoculum *Staphylococcus aureus* (both methicillin-susceptible and resistant *S. aureus*) on the activities of different antibiotics. At moderate inoculum, nafcillin, vancomycin and daptomycin were bactericidal and killed 99.9% of bacteria, whereas linezolid was bacteriostatic. At high inoculum, MIC against methicillin-susceptible *S. aureus* increased 32-fold for daptomycin and 4-fold for nafcillin and vancomycin, whereas it remained unchanged for linezolid. This was true for methicillin-resistant *S. aureus* also, except for nafcillin, which showed more than a 32-fold increase in MIC. These results were almost consistent with Udekwu *et al.* (2009). However, a 32-fold increase in MIC due to high inoculum had only a minimal effect on the bactericidal activity of daptomycin (LaPlante and Rybak, 2004). Thus, high inoculum had an impact only on nafcillin and vancomycin, whereas daptomycin was affected minimally and no effect was noticed on the activity of linezolid.

Even though the phenomenon of the inoculum effect can be seen in *in vitro* tests, its *in vivo* significance is still debated. Based on a positive inoculum effect, some researchers argue that, when using t>MIC as the parameter for predicting the efficacy of β-lactams, the correlation will be better if MIC obtained from a large inoculum is used rather than using MIC obtained from the standard inoculum (Nannini *et al.* 2009; Soriano *et al.* 1996b; Soriano *et al.* 1997; Soriano and Ponte 2009). Soriano and Ponte (2009) warn that the treated animals may die if the MIC from the standard inoculum is used. Similarly, Nannini *et al.* (2009) concluded that the inoculum effect with cefazolin might result in the treatment failure in infections caused by MSSA. This conclusion was based on the study on three of the six subjects with MSSA isolates producing β-lactamase typeA who failed cefazolin therapy and on six other patients from whom no MSSA strains were isolated but had successful treatment with cefazolin. However, the finding may not have statistical significance as suggested by the authors themselves and it is not known whether the six subjects who failed cefazolin therapy had maintained sufficient t>MIC, the most important predictor of efficacy. In another experiment, an *in vitro* susceptibility test with t>MIC calculated using low inoculum of ESBL-producing *Klebsiella pneumoniae* was shown to be predictive of *in vivo* outcomes for amikacin and imipenem but not for cefepime (Szabo *et al.* 2001). Similarly, the activities of nafcillin and vancomycin were significantly reduced by a high inoculum of *S. aureus* whereas the inoculum least affected daptomycin and linezolid activities (LaPlante and Rybak 2004).

On the other hand, many researchers consider the inoculum effect to be only a laboratory phenomenon and an artifact and support the use of conventional MIC values for the PK/PD assessment of antibiotics (Craig *et al.* 2004; Brook 1989; Maglio *et al.* 2004; Bhavnani *et al.* 2006). It is argued that *in vitro* inoculum effect is only due to the effects of hydrolyzing enzymes produced by bacteria, which may increase *in vitro* MIC but may not have significance in clinical settings where the dosing regimens are repeated for days (Craig *et al.* 2004). For both low and high inocula of ESBL-producing *E. coli*, the same dose of cefepime produced similar reductions in bacterial density when t>MIC exceeded 70% (Maglio *et al.* 2004). Similarly, no significant differences were noticed between ESBL and non-ESBL producing *E. coli* in the magnitude of kill when t>MIC was 70% (Maglio *et al.* 2004). In a rabbit model of endophthalmitis, cefopera-zone and imipenem were found to be effective in reducing the bacterial density of a high inoculum of *Klebsiella pneumoniae* even though MIC was calculated using the standard inoculum size (Davey and Barza 1987). Similarly, significant bacterial killing of three ESBL isolates of *E. coli* was noticed when t>MIC of 30-45% and 20-30% was achieved for meropenem and ertapenem, respectively (DeRyke *et al.* 2007). In this case also, MIC was calculated using only the stand-ard inoculum size even though a 64-fold increase in MIC was noticed with a high inoculum size.

Post-antibiotic effect

Post-antibiotic effect (PAE) refers to the continued suppression of bacterial growth even after the removal of antibiotics following an initial exposure of bacteria to antibiotics (McDonald *et al.* 1977; Bundtzen *et al.* 1981; MacKenzie and Gould 1993). Thus, it is the time taken by a bacterial culture exposed to antibiotics to resume growth after the removal of the antibiotic from the medium. The exact mechanisms for the phenomenon of PAE are not clear. It can be due to sublethal damage to the organism due to the exposure to antibiotics thus requiring more time for repair (den Hollander *et al.* 1998), the persistence of antibiotics retained in the cell which may still exert an inhibitory effect (Stubbings *et al.* 2005; Stubbings *et al.* 2006), or the emergence of phenotypically resistant subpopulations (den Hollander *et al.* 1996). den Hollander *et al.* (1998) argued that intra-bacterial antibiotics may only have a small effect and that the PAE could be predominantly due to sublethal damage in the bacteria, resulting in decreased DNA, RNA or protein synthesis. The recovery from aminoglycoside-induced PAE may correspond to the recovery of protein synthesis (Barmada *et al.* 1993; Stubbings *et al.* 2006), whereas the recovery from quinolone (Guan *et al.* 1992) and rifampicin (Stubbings *et al.* 2006) induced PAE may correspond to the recovery of DNA and RNA synthesis, respectively.

The duration of PAE is influenced by the type of antibiotics and organism, concentration of antibiotics, total time of exposure, inoculum size, antibiotic combinations etc. (Bermudez *et al.* 1992; Zhanel *et al.* 1991; Bundtzen *et al.* 1981; Gudmundsson *et al.* 1986; Bustamante *et al.* 1984; Hanberger *et al.* 1990; den Hollander *et al.* 1996; Burgess 1999). In general, antibiotics that inhibit protein or DNA synthesis exhibit a longer PAE than those acting on the cell wall (Burgess 1999). β-lactam antibiotics exhibit a PAE against gram-positive organisms but produce only a negligible PAE against gram-negative organisms (Turnidge 1998; Gudmundsson *et al.* 1986; Hassan *et al.* 1998). A probable exception to this is the carbapenems, which produce a PAE against gram-negative organisms also (Bustamante *et al.* 1984; Gudmundsson *et al.* 1986).

PAE is calculated using the formula PAE= T-C, where T is the time taken by the antibiotic-exposed bacterial culture to increase the count to $1\log_{10}$ after the removal of the antibiotic and C is the time taken for the control, not treated by antibiotics (Dominguez *et al.* 2001; Odenholt 2001). Once the bacterial culture resumes its growth, the growth rate is the same in the control and in the treated culture; thus the exponential phases of both cultures run parallel to one another (Dominguez *et al.* 2001). Traditionally, viable counts on agar are used to calculate the PAE, but other techniques including spectrophotometry (Dominguez *et al.* 2001), bioluminescence (Hanberger *et al.* 1993), and radiometry (Fuursted 1997) are also used.

There is lack of specific guidelines in calculating PAE; thus differences in experimental methods, concentration of antibiotics used, total time of exposure and inoculum size may affect the duration of PAE in individual experiments. For example, concentration of antibiotics as multiples of MIC used to determine PAE in different experiments were 1xMIC (Speciale *et al.* 1995), 2xMIC (Lavigne *et al.* 2004), 5xMIC (Drabu and Blakemore 1990; Stubbings *et al.* 2006) or 10xMIC (Athamna *et al.* 2004; Odenholt *et al.* 2001).

Differences in PAE can be noticed between different pathogen-antibiotic combinations (Drabu and Blakemore 1990; Alados *et al.* 1990; Gudmundsson *et al.* 1986; McGrath *et al.* 1993; Bustamante *et al.* 1984). Against *Mycobacterium*, antibiotics exhibit a wide range of PAE, from less than 1h with moxifloxacin to more than 100 h with rifapentine (Chan *et al.* 2004). Similarly, both ciprofloxacin and norfloxacin, but not nalidixic acid, produced a PAE with urinary isolates of *E. coli* (Alados *et al.* 1990). Following three consecutive 2 h exposures of *Pseudomonas aeruginosa* to imipenem, temafloxacin and tobramycin, PAE was consistently seen with imipenem, increased with repeated exposure to temafloxacin, and completely disappeared by the third exposure to tobramycin (McGrath *et al.* 1993). A consistent PAE against four strains of *Pseudomonas aeruginosa* was seen with imipenem, a β-lactam antibiotic, but not with ceftazidime, another β-lactam antibiotic (Bustamante *et al.* 1984). This was consistent with the findings of Gudmundsson *et al.* (1986), where imipenem but not cefoperazone produced a PAE against *P. aeruginosa*.

PAE can be observed *in vivo* also (Renneberg and Walder 1989; Safdar *et al.* 2004; Fantin *et al.* 1991*a*, *b*; Craig 1993) and is usually longer than *in vitro* PAE (Craig 1993; Renneberg and Walder 1989; Tauber *et al.* 1984). The longer *in vivo* PAE is attributed to the presence of neutrophils (Craig 1993). The duration of *in vivo* PAE was found to have no correlation with MIC or *in vitro* PAE (Fantin *et al.* 1991a). However, other findings show a correlation between *in vitro* and *in vivo* PAEs (Renneberg and Walder 1989; Vogelman *et al.* 1988). *In vivo* PAE was found to be longer in normal mice than in neutropenic mice but shorter in mice with normal renal function than in mice with renal impairment (Fantin *et al.* 1991a).

The knowledge about PAE may be helpful in designing the dosage regimens of antibiotics. For those antibiotics that produce longer PAE, like aminoglycosides and fluoroquinolones, sub-MIC levels can be allowed towards the end of dosing intervals (Craig 1993; Pea and Viale 2006) since bacteria will be suppressed from regrowth during PAE even though the antibiotic concentration remains under MIC. Thus, a prolonged PAE may allow once daily dosing of aminoglycosides (Gilbert 1991; Isaksson *et al.* 1988; Craig 1993). For β-lactam antibiotics that show negligible PAE, sub-MIC levels may allow bacteria to regrow and hence wider dosing intervals cannot be allowed (Pea and Viale 2006). However, other findings question the clinical importance of PAE (den Hollander *et al.* 1998). It was argued that, even if it has some clinical relevance, it could possibly be due to sub-MIC effect only (den Hollander *et al.* 1998). Along similar lines, Odenholt (2001) argued that, since suprainhibitory concentrations of antibiotics are always followed by sub-MIC levels *in vivo*, the post-antibiotic sub-MIC effect (PA-SME) or post-MIC effect (PME) would be more clinically significant than PAE itself.

Conclusion

Even though PD parameters such as MIC or MBC are good predictors of the potency of an antibiotic, they are not adequate to predict clinical outcomes because they do not give any indication about the time course of antibacterial activity nor can they account for the fluctuations in the concentration of antibiotics within the body. Moreover, they do not provide any information on the antibiotic activity and interacton of the drug at the site of infection. However, PK/PD parameters such as AUC/MIC, t>MIC and Cmax/MIC can provide a much better prediction for *in vivo* antibacterial activity. The knowledge on PK/PD allows the selection of optimal dosage regimen and determines clinically relevant susceptibility breakpoints and thus helps to predict the clinical outcome.

References

Alados, J. C., Gutierrez, J., Garcia, F., Liebana, J., and Piedrola, G. (1990). Postantibiotic effect of three quinolones against gram negative isolates from urine. *Med Lab Sci* 47(4), 272-7.

Almeida, D., Nuermberger, E., Tyagi, S., Bishai, W. R., and Grosset, J. (2007). *In vivo* validation of the mutant selection window hypothesis with moxifloxacin in a murine model of tuberculosis. *Antimicrob Agents Chemother* 51(12), 4261-6.

Ambrose, P. G., Bhavnani, S. M., Rubino, C. M., Louie, A., Gumbo, T., Forrest, A., and Drusano, G. L. (2007). Pharmacokinetics-pharmacodynamics of antimicrobial therapy: it's not just for mice anymore. *Clin Infect Dis* 44(1), 79-86.

Ambrose, P. G., Grasela, D. M., Grasela, T. H., Passarell, J., Mayer, H. B., and Pierce, P. F. (2001). Pharmacodynamics of fluoroquinolones against Streptococcus pneumoniae in patients with community-acquired respiratory tract infections. *Antimicrob Agents Chemother* 45(10), 2793-7.

Athamna, A., Athamna, M., Medlej, B., Bast, D. J., and Rubinstein, E. (2004). *In vitro* post-antibiotic effect of fluoroquinolones, macrolides, beta-lactams, tetracyclines, vancomycin, clindamycin, linezolid, chloramphenicol, quinupristin/dalfopristin and rifampicin on Bacillus anthracis. *J Antimicrob Chemother* 53(4), 609-15.

Balko, T., Karlowsky, J. A., Palatnick, L. P., Zhanel, G. G., and Hoban, D. J. (1999). Characterization of the inoculum effect with Haemophilus influenzae and beta-lactams. *Diagn Microbiol Infect Dis* 33(1), 47-58.

Barmada, S., Kohlhepp, S., Leggett, J., Dworkin, R., and Gilbert, D. (1993). Correlation of tobramycin-induced inhibition of protein synthesis with postantibiotic effect in Escherichia coli. *Antimicrob Agents Chemother* 37(12), 2678-83.

Benner, E. J., Bennett, J. V., Brodie, J. L., and Kirby, W. M. M. (1965). Inactivation of Cephalothin and Cephaloridine by Staphylococcus Aureus. *Journal of Bacteriology* 90(6), 1599-1604.

Bermudez, L. E., Wu, M., Young, L. S., and Inderlied, C. B. (1992). Postantibiotic effect of amikacin and rifapentine against Mycobacterium avium complex. *J Infect Dis* 166(4), 923-6.

Bhavnani, S. M., Ambrose, P. G., Craig, W. A., Dudley, M. N., and Jones, R. N. (2006). Outcomes evaluation of patients with ESBL- and non-ESBL-producing Escherichia coli and Klebsiella species as defined by CLSI reference methods: report from the SENTRY Antimicrobial Surveillance Program. *Diagn Microbiol Infect Dis* 54(3), 231-6.

Brook, I. (1989). Inoculum effect. *Rev Infect Dis* 11(3), 361-8.

Bundtzen, R. W., Gerber, A. U., Cohn, D. L., and Craig, W. A. (1981). Postantibiotic suppression of bacterial growth. *Rev Infect Dis* 3(1), 28-37.

Burgess, D. S. (1999). Pharmacodynamic principles of antimicrobial therapy in the prevention of resistance. *Chest* 115(3 Suppl), 19S-23S.

Burgess, D. S., and Hall, R. G., 2nd (2004). *In vitro* killing of parenteral beta-lactams against standard and high inocula of extended-spectrum beta-lactamase and non-ESBL producing Klebsiella pneumoniae. *Diagn Microbiol Infect Dis* 49(1), 41-6.

Bustamante, C. I., Drusano, G. L., Tatem, B. A., and Standiford, H. C. (1984). Postantibiotic effect of imipenem on Pseudomonas aeruginosa. *Antimicrob Agents Chemother* 26(5), 678-82.

Campion, J. J., McNamara, P. J., and Evans, M. E. (2004). Evolution of ciprofloxacin-resistant Staphylococcus aureus in *in vitro* pharmacokinetic environments. *Antimicrob Agents Chemother* 48(12), 4733-44.

Cazzola, M., and Matera, M. G. (1998). Interrelationship between pharmacokinetics and pharmacodynamics in choosing the appropriate antibiotic and the dosage regimen for treating acute exacerbations of chronic bronchitis. *Respir Med* 92(7), 895-901.

Chan, C. Y., Au-Yeang, C., Yew, W. W., Leung, C. C., and Cheng, A. F. (2004). *In vitro* postantibiotic effects of rifapentine, isoniazid, and moxifloxacin against Mycobacterium tuberculosis. *Antimicrob Agents Chemother* 48(1), 340-3.

Craig, W. A. (1993). Post-antibiotic effects in experimental infection models: relationship to in-vitro phenomena and to treatment of infections in man. *J Antimicrob Chemother* 31 Suppl D, 149-58.

Craig, W. A. (1995). Interrelationship between pharmacokinetics and pharmacodynamics in determining dosage regimens for broad-spectrum cephalosporins. *Diagn Microbiol Infect Dis* 22(1-2), 89-96.

Craig, W. A. (1998). Pharmacokinetic/pharmacodynamic parameters: rationale for antibacterial dosing of mice and men. *Clin Infect Dis* 26(1), 1-10; quiz 11-2.

Craig, W. A. (2001). Does the dose matter? *Clin Infect Dis* 33 Suppl 3, S233-7.

Craig, W. A., Bhavnani, S. M., and Ambrose, P. G. (2004). The inoculum effect: fact or artifact? *Diagn Microbiol Infect Dis* 50(4), 229-30.

Croisier, D., Etienne, M., Piroth, L., Bergoin, E., Lequeu, C., Portier, H., and Chavanet, P. (2004). *In vivo* pharmacodynamic efficacy of gatifloxacin against Streptococcus pneumoniae in an experimental model of pneumonia: impact of the low levels of fluoroquinolone resistance on the enrichment of resistant mutants. *J Antimicrob Chemother* 54(3), 640-7.

Cui, J., Liu, Y., Wang, R., Tong, W., Drlica, K., and Zhao, X. (2006a). The mutant selection window in rabbits infected with Staphylococcus aureus. *J Infect Dis* 194(11), 1601-8.

Cui, L., Iwamoto, A., Lian, J. Q., Neoh, H. M., Maruyama, T., Horikawa, Y., and Hiramatsu, K. (2006b). Novel mechanism of antibiotic resistance originating in vancomycin-intermediate Staphylococcus aureus. *Antimicrob Agents Chemother* 50(2), 428-38.

Davey, P. G., and Barza, M. (1987). The inoculum effect with gram-negative bacteria *in vitro* and *in vivo. J Antimicrob Chemother* 20(5), 639-44.

den Hollander, J. G., Fuursted, K., Verbrugh, H. A., and Mouton, J. W. (1998). Duration and clinical relevance of postantibiotic effect in relation to the dosing interval. *Antimicrob Agents Chemother* 42(4), 749-54.

den Hollander, J. G., Mouton, J. W., van Goor, M. P., Vleggaar, F. P., and Verbrugh, H. A. (1996). Alteration of postantibiotic effect during one dosing interval of tobramycin, simulated in an *in vitro* pharmacokinetic model. *Antimicrob Agents Chemother* 40(3), 784-6.

DeRyke, C. A., Banevicius, M. A., Fan, H. W., and Nicolau, D. P. (2007). Bactericidal activities of meropenem and ertapenem against extended-spectrum-beta-lactamase-producing Escherichia coli and Klebsiella pneumoniae in a neutropenic mouse thigh model. *Antimicrob Agents Chemother* 51(4), 1481-6.

Dominguez, M. C., de La Rosa, M., and Borobio, M. V. (2001). Application of a spectrophotometric method for the determination of post-antibiotic effect and comparison with viable counts in agar. *J Antimicrob Chemother* 47(4), 391-8.

Dong, Y., Zhao, X., Domagala, J., and Drlica, K. (1999). Effect of fluoroquinolone concentration on selection of resistant mutants of Mycobacterium bovis BCG and Staphylococcus aureus. *Antimicrob Agents Chemother* 43(7), 1756-8.

Drabu, Y. J., and Blakemore, P. H. (1990). Comparative post-antibiotic effect of five antibiotics against ten aerobic gram-positive cocci. *Drugs Exp Clin Res* 16(11), 557-63.

Drlica, K. (2003). The mutant selection window and antimicrobial resistance. *J Antimicrob Chemother* 52(1), 11-7.

Drlica, K., and Zhao, X. (2007). Mutant selection window hypothesis updated. *Clin Infect Dis* 44(5), 681-8.

Eng, R. H., Smith, S. M., and Cherubin, C. (1984). Inoculum effect of new beta-lactam antibiotics on Pseudomonas aeruginosa. *Antimicrob Agents Chemother* 26(1), 42-7.

Epstein, B. J., Gums, J. G., and Drlica, K. (2004). The changing face of antibiotic prescribing: the mutant selection window. *Ann Pharmacother* 38(10), 1675-82.

Fantin, B., Ebert, S., Leggett, J., Vogelman, B., and Craig, W. A. (1991a). Factors affecting duration of in-vivo postantibiotic effect for aminoglycosides against gram-negative bacilli. *J Antimicrob Chemother* 27(6), 829-36.

Fantin, B., Leggett, J., Ebert, S., and Craig, W. A. (1991b). Correlation between *in vitro* and *in vivo* activity of antimicrobial agents against gram-negative bacilli in a murine infection model. *Antimicrob Agents Chemother* 35(7), 1413-22.

Ferran, A. A., Kesteman, A. S., Toutain, P. L., and Bousquet-Melou, A. (2009). Pharmacokinetic/pharmacodynamic analysis of the influence of inoculum size on the selection of resistance in Escherichia coli by a quinolone in a mouse thigh bacterial infection model. *Antimicrob Agents Chemother* 53(8), 3384-90.

Firsov, A. A., Lubenko, I. Y., Smirnova, M. V., Strukova, E. N., and Zinner, S. H. (2008). Enrichment of fluoroquinolone-resistant Staphylococcus aureus: oscillating ciprofloxacin concentrations simulated at the upper and lower portions of the mutant selection window. *Antimicrob Agents Chemother* 52(6), 1924-8.

Firsov, A. A., Vostrov, S. N., Kononenko, O. V., Zinner, S. H., and Portnoy, Y. A. (1999). Prediction of the effects of inoculum size on the antimicrobial action of trovafloxacin and ciprofloxacin against Staphylococcus aureus and Escherichia coli in an *in vitro* dynamic model. *Antimicrob Agents Chemother* 43(3), 498-502.

Fuursted, K. (1997). Evaluation of the post-antibiotic effect of six anti-mycobacterial agents against Mycobacterium avium by the Bactec radiometric method. *J Antimicrob Chemother* 40(1), 33-8.

Gilbert, D. N. (1991). Once-daily aminoglycoside therapy. *Antimicrob Agents Chemother* 35(3), 399-405.

Gluzman, I. Y., Schlesinger, P. H., and Krogstad, D. J. (1987). Inoculum effect with chloroquine and Plasmodium falciparum. *Antimicrob Agents Chemother* 31(1), 32-6.

Guan, L., Blumenthal, R. M., and Burnham, J. C. (1992). Analysis of macromolecular biosynthesis to define the quinolone-induced postantibiotic effect in Escherichia coli. *Antimicrob Agents Chemother* 36(10), 2118-24.

Gudmundsson, S., Vogelman, B., and Craig, W. A. (1986). The in-vivo postantibiotic effect of imipenem and other new antimicrobials. *J Antimicrob Chemother* 18 Suppl E, 67-73.

Hanberger, H., Nilsson, L. E., Kihlstrom, E., and Maller, R. (1990). Postantibiotic effect of beta-lactam antibiotics on Escherichia coli evaluated by bioluminescence assay of bacterial ATP. *Antimicrob Agents Chemother* 34(1), 102-6.

Hanberger, H., Svensson, E., Nilsson, M., Nilsson, L. E., Hornsten, E. G., and Maller, R. (1993). Effects of imipenem on Escherichia coli studied using bioluminescence, viable counting and microscopy. *J Antimicrob Chemother* 31(2), 245-60.

Hansen, G. T., Zhao, X., Drlica, K., and Blondeau, J. M. (2006). Mutant prevention concentration for ciprofloxacin and levofloxacin with Pseudomonas aeruginosa. *Int J Antimicrob Agents* 27(2), 120-4.

Hassan, I. J., Stark, R. M., Greenman, J., and Millar, M. R. (1998). Absence of a post-antibiotic effect (PAE) of beta-lactams against Helicobacter pylori NCTC 11637. *J Antimicrob Chemother* 42(5), 661-3.

Homma, T., Hori, T., Sugimori, G., and Yamano, Y. (2007). Pharmacodynamic assessment based on mutant prevention concentrations of fluoroquinolones to prevent the emergence of resistant mutants of Streptococcus pneumoniae. *Antimicrob Agents Chemother* 51(11), 3810-5.

Hovde, L. B., Rotschafer, S. E., Ibrahim, K. H., Gunderson, B., Hermsen, E. D., and Rotschafer, J. C. (2003). Mutation prevention concentration of ceftriaxone, meropenem, imipenem, and ertapenem against three strains of Streptococcus pneumoniae. *Diagn Microbiol Infect Dis* 45(4), 265-7.

Isaksson, B., Nilsson, L., Maller, R., and Soren, L. (1988). Postantibiotic effect of aminoglycosides on gram-negative bacteria evaluated by a new method. *J Antimicrob Chemother* 22(1), 23-33.

Jacobs, M. R. (2003). How can we predict bacterial eradication? *Int J Infect Dis* 7 Suppl 1, S13-20.

Kang, C. I., Pai, H., Kim, S. H., Kim, H. B., Kim, E. C., Oh, M. D., and Choe, K. W. (2004). Cefepime and the inoculum effect in tests with Klebsiella pneumoniae producing plasmid-mediated AmpC-type beta-lactamase. *J Antimicrob Chemother* 54(6), 1130-3.

Kashuba, A. D., Bertino, J. S., Jr., and Nafziger, A. N. (1998). Dosing of aminoglycosides to rapidly attain pharmacodynamic goals and hasten therapeutic response by using individualized pharmacokinetic monitoring of patients with pneumonia caused by gram-negative organisms. *Antimicrob Agents Chemother* 42(7), 1842-4.

Kobayashi, H., Takemura, Y., and Ohnuma, T. (1992). Relationship between tumor cell density and drug concentration and the cytotoxic effects of doxorubicin or vincristine: mechanism of inoculum effects. *Cancer Chemother Pharmacol* 31(1), 6-10.

Konig, C., Simmen, H. P., and Blaser, J. (1998). Bacterial concentrations in pus and infected peritoneal fluid--implications for bactericidal activity of antibiotics. *J Antimicrob Chemother* 42(2), 227-32.

LaPlante, K. L., and Rybak, M. J. (2004). Impact of high-inoculum Staphylococcus aureus on the activities of nafcillin, vancomycin, linezolid, and daptomycin, alone and in combination with gentamicin, in an *in vitro* pharmacodynamic model. *Antimicrob Agents Chemother* 48(12), 4665-72.

Lavigne, J. P., Bonnet, R., Michaux-Charachon, S., Jourdan, J., Caillon, J., and Sotto, A. (2004). Post-antibiotic and post-beta-lactamase inhibitor effects of ceftazidime plus sulbactam on extended-spectrum beta-lactamase-producing Gram-negative bacteria. *J Antimicrob Chemother* 53(4), 616-9.

Livermore, D. M. (2003). Overstretching the mutant prevention concentration. *J Antimicrob Chemother* 52(4), 732; author reply 732-3.

Lu, T., Zhao, X., Li, X., Hansen, G., Blondeau, J., and Drlica, K. (2003). Effect of chloramphenicol, erythromycin, moxifloxacin, penicillin and tetracycline concentration on the recovery of resistant mutants of Mycobacterium smegmatis and Staphylococcus aureus. *J Antimicrob Chemother* 52(1), 61-4.

MacKenzie, F. M., and Gould, I. M. (1993). The post-antibiotic effect. *J Antimicrob Chemother* 32(4), 519-37.

Maglio, D., Ong, C., Banevicius, M. A., Geng, Q., Nightingale, C. H., and Nicolau, D. P. (2004). Determination of the *in vivo* pharmacodynamic profile of cefepime against extended-spectrum-beta-lactamase-producing Escherichia coli at various inocula. *Antimicrob Agents Chemother* 48(6), 1941-7.

Mariat, C., Venet, C., Jehl, F., Mwewa, S., Lazarevic, V., Diconne, E., Fonsale, N., Carricajo, A., Guyomarc'h, S., Vermesch, R., Aubert, G., Bidault, R., Bertrand, J. C., and Zeni, F. (2006). Continuous infusion of ceftazidime in critically ill patients undergoing continuous venovenous haemodiafiltration: pharmacokinetic evaluation and dose recommendation. *Crit Care* 10(1), R26.

McDonald, P. J., Craig, W. A., and Kunin, C. M. (1977). Persistent effect of antibiotics on Staphylococcus aureus after exposure for limited periods of time. *J Infect Dis* 135(2), 217-23.

McGrath, B. J., Lamp, K. C., and Rybak, M. J. (1993). Pharmacodynamic effects of extended dosing intervals of imipenem alone and in combination with amikacin against Pseudomonas aeruginosa in an *in vitro* model. *Antimicrob Agents Chemother* 37(9), 1931-7.

Morrissey, I., and George, J. T. (1999). The effect of the inoculum size on bactericidal activity. *J Antimicrob Chemother* 43(3), 423-5.

Morrissey, I., and Smith, J. T. (1994). The importance of oxygen in the killing of bacteria by ofloxacin and ciprofloxacin. *Microbios* 79(318), 43-53.

Mouton, J. W., and Vinks, A. A. (1996). Is continuous infusion of beta-lactam antibiotics worthwhile?--efficacy and pharmacokinetic considerations. *J Antimicrob Chemother* 38(1), 5-15.

Mueller, M., de la Pena, A., and Derendorf, H. (2004). Issues in pharmacokinetics and pharmacodynamics of anti-infective agents: kill curves versus MIC. *Antimicrob Agents Chemother* 48(2), 369-77.

Nannini, E. C., Stryjewski, M. E., Singh, K. V., Bourgogne, A., Rude, T. H., Corey, G. R., Fowler, V. G., Jr., and Murray, B. E. (2009). Inoculum effect with cefazolin among clinical isolates of methicillin-susceptible Staphylococcus aureus: frequency and possible cause of cefazolin treatment failure. *Antimicrob Agents Chemother* 53(8), 3437-41.

Navas, D., Caillon, J., Batard, E., Le Conte, P., Kergueris, M. F., Moreau, P., and Potel, G. (2006). Trough serum concentrations of beta-lactam antibiotics in cancer patients: inappropriateness of conventional schedules to pharmacokinetic/pharmacodynamic properties of beta-lactams. *Int J Antimicrob Agents* 27(2), 102-7.

Odenholt, I. (2001). Pharmacodynamic effects of subinhibitory antibiotic concentrations. *Int J Antimicrob Agents* 17(1), 1-8.

Odenholt, I., Lowdin, E., and Cars, O. (2001). Pharmacodynamics of telithromycin *In vitro* against respiratory tract pathogens. *Antimicrob Agents Chemother* 45(1), 23-9.

Olofsson, S. K., Marcusson, L. L., Komp Lindgren, P., Hughes, D., and Cars, O. (2006). Selection of ciprofloxacin resistance in Escherichia coli in an *in vitro* kinetic model: relation between drug exposure and mutant prevention concentration. *J Antimicrob Chemother* 57(6), 1116-21.

Olofsson, S. K., Marcusson, L. L., Stromback, A., Hughes, D., and Cars, O. (2007). Dose-related selection of fluoroquinolone-resistant Escherichia coli. *J Antimicrob Chemother* 60(4), 795-801.

Pea, F., and Viale, P. (2006). The antimicrobial therapy puzzle: could pharmacokinetic-pharmacodynamic relationships be helpful in addressing the issue of appropriate pneumonia treatment in critically ill patients? *Clin Infect Dis* 42(12), 1764-71.

Preston, S. L., Drusano, G. L., Berman, A. L., Fowler, C. L., Chow, A. T., Dornseif, B., Reichl, V., Natarajan, J., and Corrado, M. (1998). Pharmacodynamics of levofloxacin: a new paradigm for early clinical trials. *JAMA* 279(2), 125-9.

Renneberg, J., and Walder, M. (1989). Postantibiotic effects of imipenem, norfloxacin, and amikacin *in vitro* and *in vivo*. *Antimicrob Agents Chemother* 33(10), 1714-20.

Rybak, M. J. (2006). Pharmacodynamics: relation to antimicrobial resistance. *Am J Med* 119(6 Suppl 1), S37-44; discussion S62-70.

Sabath, L. D., Garner, C., Wilcox, C., and Finland, M. (1975). Effect of inoculum and of beta-lactamase on the anti-staphylococcal activity of thirteen penicillins and cephalosporins. *Antimicrob Agents Chemother* 8(3), 344-9.

Safdar, N., Andes, D., and Craig, W. A. (2004). *In vivo* pharmacodynamic activity of daptomycin. *Antimicrob Agents Chemother* 48(1), 63-8.

Schentag, J. J. (2000). Clinical pharmacology of the fluoroquinolones: studies in human dynamic/kinetic models. *Clin Infect Dis* 31 Suppl 2, S40-4.

Sieradzki, K., and Tomasz, A. (2006). Inhibition of the autolytic system by vancomycin causes mimicry of vancomycin-intermediate Staphylococcus aureus-type resistance, cell concentration dependence of the MIC, and antibiotic tolerance in vancomycin-susceptible S. aureus. *Antimicrob Agents Chemother* 50(2), 527-33.

Smith, H. J., Nichol, K. A., Hoban, D. J., and Zhanel, G. G. (2003). Stretching the mutant prevention concentration (MPC) beyond its limits. *J Antimicrob Chemother* 51(6), 1323-5.

Soriano, F., Aguilar, L., and Ponte, C. (1997). *In vitro* antibiotic sensitivity testing breakpoints and therapeutic activity in induced infections in animal models. *J Chemother* 9 Suppl 1, 36-46.

Soriano, F., Coronel, P., Gimeno, M., Jimenez, M., Garcia-Corbeira, P., and Fernandez-Roblas, R. (1996a). Inoculum effect and bactericidal activity of cefditoren and other antibiotics against Streptococcus pneumoniae, Haemophilus influenzae, and Neisseria meningitidis. *Eur J Clin Microbiol Infect Dis* 15(9), 761-3.

Soriano, F., Garcia-Corbeira, P., Ponte, C., Fernandez-Roblas, R., and Gadea, I. (1996b). Correlation of pharmacodynamic parameters of five beta-lactam antibiotics with therapeutic efficacies in an animal model. *Antimicrob Agents Chemother* 40(12), 2686-90.

Soriano, F., and Ponte, C. (2009). Comment on: Functional relationship between bacterial cell density and the efficacy of antibiotics. *J Antimicrob Chemother* 63(6), 1301.

Soriano, F., Ponte, C., Santamaria, M., and Jimenez-Arriero, M. (1990). Relevance of the inoculum effect of antibiotics in the outcome of experimental infections caused by Escherichia coli. *J Antimicrob Chemother* 25(4), 621-7.

Speciale, A., Blandino, G., Nicolosi, V. M., Siracusa, V., and Caccamo, F. (1995). Bactericidal kinetics and postantibiotic effect of sparfloxacin against selected species of respiratory pathogens. *J Chemother* 7(6), 530-4.

Stearne, L. E., Goessens, W. H., Mouton, J. W., and Gyssens, I. C. (2007). Effect of dosing and dosing frequency on the efficacy of ceftizoxime and the emergence of ceftizoxime resistance during the early development of murine abscesses caused by Bacteroides fragilis and Enterobacter cloacae mixed infection. *Antimicrob Agents Chemother* 51(10), 3605-11.

Steels, H., James, S. A., Roberts, I. N., and Stratford, M. (2000). Sorbic acid resistance: the inoculum effect. *Yeast* 16(13), 1173-83.

Stevens, D. L., Yan, S., and Bryant, A. E. (1993). Penicillin-binding protein expression at different growth stages determines penicillin efficacy *in vitro* and *in vivo*: an explanation for the inoculum effect. *J Infect Dis* 167(6), 1401-5.

Stubbings, W., Bostock, J., Ingham, E., and Chopra, I. (2005). Deletion of the multiple-drug efflux pump AcrAB in Escherichia coli prolongs the postantibiotic effect. *Antimicrob Agents Chemother* 49(3), 1206-8.

Stubbings, W., Bostock, J., Ingham, E., and Chopra, I. (2006). Mechanisms of the post-antibiotic effects induced by rifampicin and gentamicin in Escherichia coli. *J Antimicrob Chemother* 58(2), 444-8.

Szabo, D., Mathe, A., Filetoth, Z., Anderlik, P., Rokusz, L., and Rozgonyi, F. (2001). *In vitro* and *in vivo* activities of amikacin, cefepime, amikacin plus cefepime, and imipenem against an SHV-5 extended-spectrum beta-lactamase-producing Klebsiella pneumoniae strain. *Antimicrob Agents Chemother* 45(4), 1287-91.

Takemura, Y., Kobayashi, H., Miyachi, H., Hayashi, K., Sekiguchi, S., and Ohnuma, T. (1991a). The influence of tumor cell density on cellular accumulation of doxorubicin or cisplatin *in vitro*. *Cancer Chemother Pharmacol* 27(6), 417-22.

Takemura, Y., Ohnuma, T., and Sekiguchi, S. (1991b). Antitumor efficacy of doxorubicin in combination with cisplatin on human lymphoma cells at various cell densities *in vitro*. *Keio J Med* 40(2), 78-81.

Tam, V. H., Ledesma, K. R., Chang, K. T., Wang, T. Y., and Quinn, J. P. (2009). Killing of Escherichia coli by beta-lactams at different inocula. *Diagn Microbiol Infect Dis* 64(2), 166-71.

Tauber, M. G., Zak, O., Scheld, W. M., Hengstler, B., and Sande, M. A. (1984). The postantibiotic effect in the treatment of experimental meningitis caused by Streptococcus pneumoniae in rabbits. *J Infect Dis* 149(4), 575-83.

Thomson, K. S., and Moland, E. S. (2001). Cefepime, piperacillin-tazobactam, and the inoculum effect in tests with extended-spectrum beta-lactamase-producing Enterobacteriaceae. *Antimicrob Agents Chemother* 45(12), 3548-54.

Turnidge, J. D. (1998). The pharmacodynamics of beta-lactams. *Clin Infect Dis* 27(1), 10-22.

Udekwu, K. I., Parrish, N., Ankomah, P., Baquero, F., and Levin, B. R. (2009). Functional relationship between bacterial cell density and the efficacy of antibiotics. *J Antimicrob Chemother* 63(4), 745-57.

Van Bambeke, F., and Tulkens, P. M. (2001). Macrolides: pharmacokinetics and pharmacodynamics. *Int J Antimicrob Agents* 18 Suppl 1, S17-23.

van Zanten, A. R., Oudijk, M., Nohlmans-Paulssen, M. K., van der Meer, Y. G., Girbes, A. R., and Polderman, K. H. (2007). Continuous vs. intermittent cefotaxime administration in patients with chronic obstructive pulmonary disease and respiratory tract infections: pharmacokinetics/pharmacodynamics, bacterial susceptibility and clinical efficacy. *Br J Clin Pharmacol* 63(1), 100-9.

Vogelman, B., Gudmundsson, S., Turnidge, J., Leggett, J., and Craig, W. A. (1988). *In vivo* postantibiotic effect in a thigh infection in neutropenic mice. *J Infect Dis* 157(2), 287-98.

Woodnutt, G. (2000). Pharmacodynamics to combat resistance. *J Antimicrob Chemother* 46 Suppl T1, 25-31.

Yanagisawa, C., Hanaki, H., Matsui, H., Ikeda, S., Nakae, T., and Sunakawa, K. (2009). Rapid depletion of free vancomycin in medium in the presence of beta-lactam antibiotics and growth restoration in Staphylococcus aureus strains with beta-lactam-induced vancomycin resistance. *Antimicrob Agents Chemother* 53(1), 63-8.

Zhanel, G. G., Hoban, D. J., and Harding, G. K. M. (1991). The Postantibiotic Effect - a Review of *In vitro* and *In vivo* Data. *Dicp-the Annals of Pharmacotherapy* 25(2), 153-163.

Zhao, X. (2003). Clarification of MPC and the mutant selection window concept. *J Antimicrob Chemother* 52(4), 731; author reply 732-3.

Zhao, X., and Drlica, K. (2002). Restricting the selection of antibiotic-resistant mutant bacteria: measurement and potential use of the mutant selection window. *J Infect Dis* 185(4), 561-5.

Zhao, X., and Drlica, K. (2008). A unified anti-mutant dosing strategy. *J Antimicrob Chemother* 62(3), 434-6.

CHAPTER II

PERSISTERS, PHENOTYPIC SHIFT, AND CHRONIC INFECTIONS

Section 1
Persisters: A Review of the Literature

Complete sterilization of a bacterial culture is expected when an antibiotic-sensitive bacterial population is treated with optimal concentrations of antibiotics. However, Joseph Bigger noticed that penicillin could not sterilize a Staphylococcal culture completely even in the absence of a penicillin-resistant population (Bigger 1944). While most of the bacteria were killed by penicillin, Bigger noticed that a small subpopulation somehow survived which were neither killed nor grown in the presence of antibiotics. Upon removal of penicillin, the survivors grew abundantly just like the parent population. They were as sensitive as the parent population to the bactericidal action of the antibiotic but again left a small percentage of survivors. It is argued that the survival of this small subpopulation of persisters in the presence of antibiotics is not due to antibiotic resistance, but rather to their ability to undergo a phenotypic shift by remaining in a dormant, non-dividing state that enable them to survive the lethal actions of antibiotics (Bigger 1944; Lewis 2007; Kussell et al. 2005).

Bactericidal compounds typically exhibit a biphasic killing in which an initial rapid killing of bacteria is followed by a slower phase of killing (Keren et al. 2004a). The slow rate of killing is attributed to the presence of persisters. Persisters are not induced by antibiotics nor are they mutants. Their formation depends on the growth stage of bacteria (Keren et al. 2004a). During the lag and the early exponential phases of bacterial growth, the number of persisters is small. They start to increase in number during the mid exponential phase and reach a plateau by the late exponential phase. Since they are not formed in the early logarithmic phase, persisters are not cells at a particular stage in the cell cycle. Persisters are pre-formed rather than being generated during the antibiotic treatment (Keren et al. 2004a). When bacteria were continuously grown in an early exponential phase by repeated dilution and re-growth followed by antibiotic treatment, persisters disappeared, indicating that they were not formed in response to antibiotics (Keren et al. 2004a).

Isolating persisters is difficult because of their low number in the presence of antibiotics and because they undergo a phenotypic shift upon the removal of the antibiotic (Lewis 2007). However, high persister (*hiP*) mutants of *E. coli*, first isolated by Moyed and Bertrand (1983), can be used to obtain large number of persisters. After several cycles of treatment of an *E. coli* population in the presence of ampicillin followed by growing the survivors in the absence of the antibiotic, they isolated two types of mutant colonies: one with an antibiotic-resistant gene that allowed bacteria to grow in the presence of antibiotics, and one that produced a high number of persister cells that did not grow in the presence of antibiotics (Moyed and Bertrand 1983). The latter group was found to be *hipA* mutants that produced approximately 1000 times more persisters than the wild type bacteria. Persisters are tolerant to antibiotics and are neither killed nor grown in the presence of antibiotics. They exhibit increased MBC (minimum bactericidal concentration) but show almost the same MIC (minimum inhibitory concentration) as the wild type (Keren *et al.* 2004a).

A relatively easy method to isolate persisters is to treat bacterial cultures with an antibiotic for 3 h and collect the survivors by centrifugation (Keren *et al.* 2004b). Another method is to sort out bacterial cells in which green fluorescent protein (GFP) is expressed from a promoter whose activity depends on the growth rate (Shah *et al.* 2006). A majority of the exponential phase bacteria may show bright fluorescence, whereas a small subpopulation may show low or undetectable fluorescence. Based on their fluorescence intensities, the two populations can be separated. Since the persisters are dormant cells with low translational levels, GFP expression will also be low and hence the persisters (dim cells) can be separated from the normal population using a cell sorter (Shah *et al.* 2006).

During the stationary phase, the number of persisters increases and hence can be detected more readily (Roostalu *et al.* 2008). An *E. coli* population that had been in the stationary phase for 4 days may not undergo cell division (Roostalu *et al.* 2008). However, when transferred to fresh medium, they may generate two subpopulations: one that grows normally and one that remains dormant. Using the GFP dilution method and cell sorting, it was shown that treatment with ampicillin resulted in the lysis of growing bacteria only, whereas the non-growing population (persisters) was not killed and growth was restarted only when the antibiotic was removed from the medium (Roostalu *et al.* 2008).

Using a microfluidic device, Balaban *et al.* (2004) showed that persisters can be of two types: Type-I and Type-II persisters. Type-I persisters are a pre-existing population of non-growing cells produced during the stationary phase and take a longer time to exit the stationary phase (Balaban *et al.* 2004). The switching rate of normal cells to Type-I persisters during the exponential phase is negligible. Once transferred to fresh medium, they remain in a growth-arrested state for many hours. Only after remaining in the fresh medium for many hours can they switch back to normal cells (Balaban *et al.* 2004; Gefen *et al.* 2008). Type-II persisters, on the other hand, are not in a growth-arrested state but constitute slow-growing cells. They are generated continuously, and their number is determined by the total number of cells. Their formation is not dependent on a starvation signal or stationary phase (Balaban *et al.* 2004; Gefen *et al.* 2008). They are formed by a phenotype-switching mechanism wherein a normal cell spontaneously becomes a Type-II persister and vice versa (Kussell *et al.* 2005). A wild type population thus consists of three subpopulations: normal cells that grow fast and are quickly killed by antibiotics; Type-I persisters that are generated during the stationary phase of the previous cycle; and Type-II persisters that are generated continuously. Both Type-I and Type-II persisters can avoid being killed by antibiotics.

Recently, it was found that even the persisters are vulnerable to the bactericidal action of ampicillin over a narrow time window (Gefen *et al.* 2008). Type-I persisters, generated during the stationary phase, are tolerant to antibiotics; however, when transferred to fresh medium, they exit the stationary phase, during which time protein production occurs and the cells become vulnerable to antibiotics. However, this period is relatively short (approximately 1.5 h). After this short time window, protein production stops and the cells are differentiated into the dormant state to become persisters and again remain tolerant to antibiotics. Knowledge about this vulnerable period may be useful for developing new strategies against persistent bacterial infections (Gefen *et al.* 2008).

Difference between resistance, tolerance, and persistence

Resistant bacteria exhibit increased MIC to antibiotics and carry antibiotic-resistant genes that are passed on to the next generation so that the whole population becomes resistant to the antibiotic. Antibiotic tolerance, on the other hand, is characterized by an increased MBC but organisms exhibit the same MIC as the wild type. In this case, the concentration of the antibiotic needed to inhibit the growth of the bacteria is not changed, but a higher concentration is needed to kill them. *hipA* mutants, for example, exhibit antibiotic tolerance wherein the number of survivors is much larger than that of the wild type at the given MBC value (Keren *et al.* 2004a). However, tolerance may not be always due to mutations. For example, a bacterial culture may exhibit tolerance during the stationary phase. Here, the entire stationary phase culture may remain tolerant. This tolerance is not due to mutations because, once transferred to fresh medium, they become susceptible to antibiotics. Persisters, on the other hand, are non-mutants but comprise only a small subpopulation (less than 1%) of the bacterial culture. In this case, they survive even when the majority of the population is killed. They also exhibit increased MBC but show the same MIC as the wild type.

Number of persisters

Persisters represent a very small subpopulation of less than 1% of the total population (Spoering and Lewis 2001; Keren *et al.* 2004a). An initial inoculum size of 10^5 or 10^6 bacteria/ml may not produce survivors after antibiotic treatment. However, as the population enters the stationary phase, the total number of persisters reach approximately 1% of the total population (Spoering and Lewis 2001; Keren *et al.* 2004a). Because of their low numbers, persisters are usually ignored in the conventional tests that determine the MIC or MBC (Keren *et al.* 2004a; Lewis 2001). One reason is that, in those conventional tests, the inoculum size used is 10^5-10^6 bacteria per ml (at this inoculum size, persisters may be absent). Moreover, only logarithmic phase cultures are used in these tests (most of the persisters are formed in the stationary phase). Additionally, MBC is defined as the concentration that kills 99.9% of the population. Hence, even if a small subpopulation of 0.1% survives, they will not be reported as it falls within the accepted range (Keren *et al.* 2004a).

The number of persisters differs with the phase of growth, type of antibiotics and the type of bacterial community (Spoering and Lewis 2001). Logarithmic and stationary phase planktonic cultures and biofilms of *Pseudomonas aeroginosa* behaved differently with different antibiotics as far as the percentage of persisters is concerned (Spoering and Lewis 2001). Carbenicillin at 600μg/ml, treated for 6 h, resulted in the survival of 0.1% persisters, whereas ofloxacin (15μg/ml) resulted in 0.0001% persisters. Treatment with tobramycin, on the other hand, completely killed all bacteria even at 4μg/ml. The ability of ofloxacin (fluoroquinolone) and tobramycin (aminoglycoside) to kill persisters was attributed to their ability to kill non-dividing cells. However, even those antibiotics are less effective in killing persisters in the stationary phase cultures and in biofilms (Spoering and Lewis 2001). The percentage of persisters after ofloxacin treatment was 0.1% in the biofilm and 2.5% in the stationary phase cultures. Carbenicillin showed only negligible killing of stationary phase bacteria and biofilms, whereas tobramycin was ineffective in killing both the stationary phase cultures and the biofilms (Spoering and Lewis 2001).

Persisters: a more generalized phenomenon

Nitzan *et al.* (1987) reported on the antibacterial activity of hemin on *Staphylococcus aureus*. Hemin, at low concentration (3-10 μg/ml), stopped bacterial growth completely within 30 min, and cell viability was reduced by 99.9% within 1 h of exposure. The remaining bacteria, however, could regrow in fresh medium and the recovered population was found to be sensitive to hemin similar to the parent culture. Thus, the survivors were not hemin-resistant mutants, and the authors suggested that the sensitivity of bacteria to hemin was regulated by an "on-off" mechanism. The ability of bacteria to survive hemin is thus similar to the ability of persisters to survive antibiotic treatment. Persisters can also be detected after treating *E. coli* cultures with metal oxyanions (Harrison *et al.* 2005).

Anti-fungal tolerant persister cells are also reported (LaFleur *et al.* 2006; LaFleur *et al.* 2010). *Candida albicans* biofilms exhibited a biphasic killing pattern against amphotericin B, an antifungal agent, and chlorhexidine, an antiseptic, similar to the response of a bacterial culture to antibiotics, indicating the presence of fungal persisters (LaFleur *et al.* 2006). Persisters were detected only in biofilms but not in the exponential or stationary phase planktonic *C. albicans*. Those persisters exhibited multidrug tolerance, and no additional killing was detected when both the agents were used together (LaFleur *et al.* 2006). Similarly, *C. albicans* and *C. glabrata* persisters were isolated from cancer patients treated with topical chlorhexidine for many days and who were at the risk of oral candidiasis (LaFleur *et al.* 2010). However, Al-Dhaheri and Douglas (2008) reported that the persisters alone cannot explain the drug tolerance in *Candida* species. No persister cells were detected in any of the exponential or stationary phase planktonic cells of five *Candida* species and in the biofilms of *C. glabrata* or *C. tropicalis*. However, they were detected in the biofilms of *C. albicans*, *C. krusei* and *C. parapsilosis* (Al-Dhaheri and Douglas 2008), which also showed signs of apoptosis when the biofilms were exposed to amphotrecin B (Al-Dhaheri and Douglas 2010).

Significance of persisters

Many chronic infections like cystic fibrosis, osteomyelitis, tuberculosis and syphilis are difficult to treat with conventional antibiotic therapy (Young *et al.* 2002; Lewis 2007). Similarly, infections caused by biofilms may not respond well to the usual antibiotic therapy (reviewed by Costerton *et al.* 1999; Mah and O'Toole 2001; del Pozo and Patel 2007; Anderson and O'Toole 2008; Stewart and Franklin 2008). However, planktonic cells derived from these biofilms may remain fully susceptible to antibiotics, indicating that it is the biofilm itself that is responsible for the resistance (Nickel *et al.* 1985). Biofilms are formed by a dense extracellular matrix of excreted polymeric substances and are responsible for recalcitrant infections. The biofilm resistance to antibiotics can be due to reduced penetration of antibiotics into the biofilm, reduced growth of bacteria inside the biofilm, presence of phenotypically distinct cells and increased expression of multi-drug-resistant (MDR) pumps by the bacteria (Costerton *et al.* 1999; Mah and O'Toole 2001; Anderson and O'Toole 2008). Since the antibiotics need to traverse through the thick biofilm matrix, all the bacteria inside the biofilm may not be exposed to a sufficient concentration of antibiotics. Antibiotics may kill the outer layer of bacteria in the biofilm, but may leave a subpopulation of bacteria deep inside it (Costerton *et al.* 1999; Anderson and O'Toole 2008). Similarly, antibiotics may get trapped in the matrix, thus reducing the concentration of antibiotics inside (Anderson and O'Toole 2008). At the same time, other studies have shown that antibiotics can freely diffuse through the biofilms (Darouiche *et al.* 1994; Stone *et al.* 2002; del Pozo and Patel 2007; Patel 2005). Another reason for the ineffectiveness of antibiotics against biofilms is the reduced growth of bacteria inside the biofilm (Costerton *et al.* 1999; Anderl *et al.* 2003; Anderson and O'Toole 2008). It has been noticed that, deep inside the biofilms, pockets of hypoxia occur where bacteria are exposed to low oxygen levels (Anderson and O'Toole 2008). Oxygen or nutrient limitation can affect the growth rate of bacteria and since the antibiotics are less effective against slow-growing bacteria, those bacteria inside the biofilm may not respond well to antibiotics. However, ofloxacin, a fluoroquinolone, which can kill non-growing cells, cannot eliminate persisters completely, indicating that the slow growth of bacteria alone can not explain bacterial tolerance (Spoering and Lewis 2001). Yet another reason for the tolerance is the increased expression of certain proteins, especially the stress response proteins, in the biofilms (Anderson and O'Toole 2008). The stationary phase sigma factor, rpoS, and the stress response chaperones, *groES* and *dnaK*, are found to be upregulated in biofilms (Schembri *et al.* 2003; Anderson and O'Toole 2008). Stress responses may affect the biofilm bacterial resistance. Similarly, some cells in the biofilm may adopt a distinct and protective biofilm phenotype which can be a biologically programmed response (Costerton *et al.* 1999; del Pozo and Patel 2007; Stewart and Franklin 2008) so that the survival of at least a few cells can be guaranteed.

Persisters are implicated in chronic and biofilm-related infections (Lewis 2007, 2008). Persisters in the biofilm may be more important than the planktonic cells in recurrent infections (Lewis 2007, 2008). Antibiotic treatment may kill a majority of the biofilm and planktonic cells but may leave the persisters intact. Whereas the planktonic persister cells may be removed by the immune system, the biofilm persisters may be protected by the matrix from the immune cells. Once the antibiotic is removed, those persisters in the biofilm may start to grow and repopulate the biofilm, which may later release some planktonic cells causing recurrent infections. Thus, eliminating the persisters in the biofilm is an important aim in the treatment of recurrent infections (Lewis 2007, 2008). The use of antipersister drugs given in combination with conventional antibiotics, sterile surface materials, sterilizing antibiotics and pulse dosing (a high dose of antibiotic followed by a second dose when the persisters begin to grow) have been proposed to attain this aim (Lewis 2007, 2008).

Toxin-antitoxin module and persisters

The toxin-antitoxin (TA) system, found in the plasmids and chromosomes of many bacteria (Mittenhuber 1999; Pandey and Gerdes 2005), is implicated in persister formation (Spoering *et al.* 2006; Korch and Hill 2006; Keren *et al.* 2004b; Lewis 2001). This system consists of a pair of genes, one of which encodes a stable toxin and the other of which encodes an unstable antitoxin. Examples for such systems include *hipBA* (Korch and Hill 2006, Keren *et al.* 2004b), *relBE* (Gotfredsen and Gerdes 1998; Bech *et al.* 1985) and *mazEF* (Engelberg-Kulka *et al.* 2005; Metzger *et al.* 1988). In the presence of the antitoxin, the expression of the toxin, which inhibits translation, is down regulated. However, when the antitoxin level is reduced, the expression of the toxin becomes predominant, resulting in the inhibition of translation and thus the protein synthesis that finally results in programmed cell death (PCD) (Sat *et al.* 2001; Mittenhuber

1999; Pandey and Gerdes 2005; Engelberg-Kulka *et al.* 2005). Many stressful conditions can prevent the expression of toxin-antitoxin modules. In such cases, even though the expression of both toxins and anti-toxins decreases, the concentration of the antitoxin goes down faster as it is unstable, leading to an increased concentration of the toxin, which results in PCD (Hazan *et al.* 2004; Engelberg-Kulka *et al.* 2005; Sat *et al.* 2001). High temperature, DNA damage and oxidative stress can result in *mazEF*-mediated cell death (Hazan *et al.* 2004). *mazEF*-mediated cell death occurs only during the logarithmic phase and not in the stationary phase (Hazan *et al.* 2004). The lack of cell death during the stationary phase can explain why cells are resistant to many stressful conditions at this point. In fact, the stationary-phase sigma factor may be responsible for this resistance since the deletion of *rpoS* results in *mazEF*-mediated cell death during the stationary growth phase (Kolodkin-Gal and Engelberg-Kulka 2009).

Even though the induction of MazF inhibits the synthesis of most of the proteins, it still allows the synthesis of small proteins, some of which are required for the death of the majority of the population, whereas others help in the survival of a small subpopulation (Amitai *et al.* 2009). MazF thus act as a regulator that controls both cell death and cell survival. Thus, stressful conditions may lead to the programmed cell death of a majority of the population whereas a small subpopulation may survive at the expense of the dead cells (Amitai *et al.* 2009).

It was shown that several antibiotics mediate cell death by *mazEF* in *E. coli* (Sat *et al.* 2001) by inhibiting the continued expression of the labile antitoxin MazE resulting in the overexpression of the toxin MazF. It was later reported that MazF does not actually kill the cells but only induces cell stasis by inhibiting translation and/or replication and that the later induction of transcription of MazE could reverse the inhibitory effects of MazF (Pederson *et al.* 2002). However, Amitai *et al.* (2004) reported that the reversal of the inhibitory effects of MazF by MazE occur only over a short period of time. They observed that during the first 6 h of MazF overproduction, its inhibitory effects could be reversed completely by a subsequent overproduction of MazE. However, when MazF was overproduced for 8 h, the ability of MazE to reverse the effects of MazF was significantly reduced. It was argued that the cells had reached a point of no return and had suffered enough damage by MazF such that it could not be reversed by MazE.

The major TA system involved in persister cell formation is *hipBA* (Moyed and Bertrand 1983; Moyed and Broderick 1986). Mutations in *hipA* produce a high frequency of persisters (Moyed and Bertrand 1983). Similarly, overexpression of HipA results in 10-to-1000 fold increases in persister formation (Falla and Chopra 1998; Keren *et al.* 2004a; Korch and Hill 2006). In addition, overexpression of RelE, the toxin protein of another TA system, also results in a high frequency of persisters (Keren *et al.* 2004a). Thus, specific roles for these toxins in persister formation were proposed. However, not all the TA systems function in the same way. Several TA systems present in bacteria behave differently with respect to cell death (Kolodkin-Gal *et al.* 2009). While *mazEF* mediates cell death in both liquid culture and biofilms, *relBE* is involved only in liquid cultures but not in the biofilms. *chpBIK* may function as a backup system for *mazEF*. Whereas *yefM-yoeB* mediates cell death only in some cases in liquid media and not at all in biofilm formation, *dinJ-yafQ* is involved in biofilm formation, but not in the liquid culture (Kolodkin-Gal *et al.* 2009).

The importance of TA modules in persister formation is also questioned by some researchers. Vazquez-Laslop *et al.* (2006) found that the cells overexpressing proteins unrelated to TA modules also resulted in a high frequency of persisters. They found that the proteins that are toxic to the cell, when overexpressed, would result in a high frequency of persisters, thus questioning the specific roles of *hipA* or *relB* or other TA modules in persister generation. Similarly, Tsilibaris *et al.* (2007) questioned the specific role of the chromosomally-encoded TA system in influencing bacterial fitness and competitiveness. They could not find any evidence that supports the PCD hypothesis of TA modules, nor could they find any definite advantages of the TA system in the recovery of bacteria from different stressors including nutrient limitations.

Mechanism of persister formation

Bactericidal antibiotics kill cells by corrupting the antibiotic target (Lewis 2007). When the antibiotic binds to its target, it creates a toxic product. For example, aminoglycosides produce misfolded peptides, fluoroquinolones create DNA lesions, and penicillins activate autolysins leading to peptidoglycan digestion and cell death (Lewis 2005; Lewis 2007). Antibiotic resistance functions by preventing the antibiotic from binding to its target, whereas antibiotic tolerance functions by blocking the antibiotic target, thus preventing the target corruption (Keren *et al.* 2004b; Lewis 2007). Since persisters are dormant cells, antibiotics may not be able to corrupt the function of the target molecules even though they may bind to them (Lewis 2007). Thus, persisters create a tolerant state by shutting down antibiotic targets (Lewis 2005).

Keren *et al.* (2004b) proposed that the production of persisters in a wild type population can be stochastic resulting from different protein expression levels. Some cells may express high levels of HipA, which may inactivate the antibiotic targets and may prevent corrupting the functions of target molecules, leading to tolerance. The absence of persisters in the early exponential phase is attributed to low levels of 'persister proteins' (Lewis 2007; Lewis 2008). However, when an exponential phase bacterial culture is made to overexpress HipA, the number of persisters increases (Korch and Hill 2006). Thus, persister formation could be under the control of two processes: stochastic fluctuations at the level of proteins and a regulated expression of proteins that depends on the density of the population (Lewis 2007, 2008).

High levels of HipA may increase the intracellular basal expression of the alarmone (p)ppGpp, resulting in altered gene expressions (Korch *et al.* 2003). Deletion of *relA* and *spoT* genes, which are responsible for (p)ppGpp synthesis, abolishes the persister phenotype of *hiPA7* mutant, indicating that HipA acts by increasing the alarmone levels (Korch *et al.* 2003). (p)ppGpp is a regulator of adaptive response of bacteria to starvation and other stressors. In the presence of antibiotic stress, a further increase in the production of (p)ppGpp occurs, which induces a persistent physiological state (Korch *et al.* 2003). *hipA* not only generates a high frequency of persisters but also inhibits macromolecular synthesis including DNA, RNA and protein synthesis (Korch and Hill 2006). However, their ability to generate persisters may be distinct from their ability to inhibit macromolecular synthesis (Korch and Hill 2006).

Recently, Dorr *et al.* (2009) reported an inducible mechanism of persister formation mediated by SOS response. They found that ciprofloxacin treatment killed a majority of the population except for a small fraction of persisters. Those persisters were not a pre-formed subpopulation, but were induced by antibiotic treatment, a finding contrary to previous views. They noticed that an increase in ciprofloxacin concentration from 0.02 µg/ml to 0.5 µg/ml increased the average SOS induction but reduced the number of persister cells. However, when the ciprofloxacin concentration was increased to 1 or 2 μg/ml, the persister fraction paradoxically also increased. They also found that a bacterial culture previously exposed to a sub-MIC of ciprofloxacin generated more persisters than the control group without pre-treatment. The authors argued that, if persisters were preformed, the persister fraction surviving the ciprofloxacin concentration would be the same regardless of the pretreatment. It was thus concluded that persisters are not always pre-formed but can also be induced by antibiotics and that a specific high or low level of SOS induction is required for persister formation.

It is also possible that persisters are bacteria undergoing repair. Earlier, Debbia *et al.* (2001) suggested that the number of persisters may depend on DNA repair mechanisms and that persisters could be a small fraction of bacteria that are not growing due to an active DNA repair mechanism.

Role of *hipBA* in persister formation and stationary phase survival

The toxin-antitoxin module *hipBA* is thought to play an important role in persister formation since the overexpression of HipA results in a high frequency of persisters (Falla and Chopra 1998; Korch and Hill 2006). HipA is the toxin and HipB is the antitoxin that binds to HipA. HipB also act as a transcriptional repressor of the *hipBA* operon (Black *et al.* 1994). HipA is a member of the phosphatidylinositol 3/4-kinase superfamily (Corriea *et al.* 2006; Schumacher *et al.* 2009) and its kinase activity is necessary for antibiotic tolerance since the mutants defective in the kinase activity of HipA are highly susceptible to killing by ofloxacin (Correia *et al.* 2006). These mutants, however, provided some degree of protection, indicating that the kinase activity of HipA may not be entirely responsible for persister formation (Correia *et al.* 2006).

HipA is structurally homologous to human CDK2/cyclin A kinase (Honda *et al.* 2005). One of the HipA targets is EF-Tu, which catalyzes the aminoacyl-tRNA binding to the ribosomes (Schumacher *et al.* 2009). Phosphorylation of EF-Tu by HipA prevents the aminoacyl-tRNA binding that leads to translation inhibition. HipB binds to HipA to inhibit its kinase activity. Activation of HipA can occur only if HipB is removed or degraded (Schumacher *et al.* 2009).

The *hipBA* system may also be involved in the survival of bacterial cells in the stationary phase (Kawano *et al.* 2009). Among the six major TA disruptants, only *hipBA* disruptants exhibited longer lifespans than the wild type, indicating that *hipBA* regulates cell viability during the stationary phase (Kawano *et al.* 2009). Moreover, *hipBA* disruptants showed higher macromolecular synthesis, extended lifespans under anaerobic conditions and higher resistance to H_2O_2. However, under prolonged cultivation, even HipA-expressing wild type cells are killed by oxidative stress. Thus, HipA may play an important role in the early stationary phase, when viability starts to lessen, through the inhibition of macromolecular synthesis. However, on prolonged cultivation, oxidative damage may accumulate, resulting in the killing of cells with or without *hipBA* (Kawano *et al.* 2009).

Other genes involved in persister formation

Apart from the TA system, other genes are also reported to be involved in persister cell formation. Inactivation of *phoU* resulted in higher susceptibility of bacteria to antibiotics and a defect in persister formation resulting from a hyperactive metabolic state, which indicates that PhoU is a global negative regulator (Li and Zhang 2007). Similarly, the overexpression of glycerol-3-phosphate dehydrogenase (GlpD) resulted in a high frequency of persisters, whereas the deletion of *glpD* or *plsB* (glycerol-3-phosphate acyltransferase) resulted in a reduced number of persisters, indicating that these genes are required for the maintenance of a state of persistence (Spoering *et al.* 2006). Vasquez-Laslop *et al.* (2006) reported that the overexpression of non-TA genes like *dnaJ* or *PmrC* also resulted in the a high frequency of persister formation.

Screening a library of *E. coli* knockout strains identified several mutants that are involved in persister formation. Some of the important genes involved were *dnaJ, dnaK, surA, ygfA, yigB, apaH, fis, hns, hnr* and *dksA*. A long list of genes indicates redundancy in the mechanism of persister formation (Hansen *et al.* 2008).

Conclusion

Persisters are non-mutants and exhibit antibiotic tolerance that is non-heritable and reversible. Due to their ability to undergo a phenotypic shift, persisters survive antibiotic treatment; but re-grow later on the removal of antibiotics. The phenotypic shift of persisters is reported to be under genetic control since overexpression of toxins such as HipA, RelE and MazF results in high number of persisters. They are implicated in many chronic and biofilm-associated infections. Eliminating persisters in the biofilm may be helpful in the successful treatment of many recurrent infections.

Section 2
Why is the Phenotypic Shift of Persisters
an *In Vitro* Illusion?

E ven though the phenotypic shift of persisters appears to be an attractive hypothesis, there are a number of flaws in both the experimental methods and the conclusions derived from the results of the experiments, which are discussed below.

1. Incubating a bacterial culture with antibiotics for 3-6 h is a safe bet.

In most of the experiments demonstrating persisters, bacterial cultures were incubated for 3-6 h (Spoering and Lewis 2001; Keren *et al.* 2004a, b; Kaldalu *et al.* 2004; Lewis 2005; Shah *et al.* 2006; Dorr *et al.* 2009; Balaban *et al.* 2004). At the end of the incubation period, the majority of the bacteria were found to be dead, except for a small subpopulation of persisters. A different result would have been obtained, however, had the researchers incubated the bacterial culture for 24 or 48 h (for MIC/MBC tests, NCLLS recommends 18-24 h of incubation). Incubating the bacterial culture for a short period is a safe bet as it appears to give a consistent result.

Harrison *et al.* (2005) studied the effect of antibiotics on killing planktonic and biofilm bacterial cultures. They found that MBC_{100} (the minimum concentration of antibiotics that kill 100% of bacteria) of amikacin for a planktonic culture of *E. coli JM 109* after 24 h of incubation was 20±8 µg/ml. This indicated that kanamycin at a concentration of 20±8 µg/ml could eradicate all bacteria without leaving any persisters. Similarly, the authors reported an MBC_{100} value for ceftrioxone and tobramycin for the same organism. The authors also reported that 2 h incubation with amikacin or ceftrioxone could not give an MBC_{100}, indicating that a short incubation time may not kill all the bacteria that are otherwise susceptible to antibiotics.

2. Researchers have not provided ideal *in vitro* conditions for bacterial killing.

One of the assumptions made based on the experiments is that antibiotics are not capable of killing 100% of the bacteria (Lewis 2007). However, more stringent experimental conditions are needed

MISINTERPRETATION OF SCIENTIFIC DATA

before reaching such conclusions since a few surviving bacteria, otherwise susceptible to antibiotics, can skew the results.

A number of factors need to be considered for efficient *in vitro* bacterial killing. Apart from the total time of incubation, the initial inoculum size can affect *in vitro* bacterial killing (Peterson and Shanholtzer 1992; Masuda *et al.* 1979; Brook 1989; Eng *et al.* 1985; Firsov *et al.* 1997). An 'inoculum effect' is well documented with many antibiotics (LaPlante and Rybak 2004; Eng *et al.* 1985; Bulger and Washington 1980; Konig *et al.* 1998; Davey and Barza 1987). In most cases, a 'phenotypic shift' can be attributed solely to the inoculum effect, which is discussed later. The concentration of the antibiotic used in the bacterial killing experiments is another factor that can determine the outcome. Antibacterial activity of antibiotics, in general, is proportional to its concentration. However, at very high concentrations of some antibiotics, this may not be true. Antibiotics are most effective in a narrow range of concentration above the MIC (Woolfrey *et al.* 1987, Woolfrey and Enright 1990). When the concentration of antibiotics is well above the MIC, a paradoxical effect can be noticed (Holm *et al.* 1990; Crumplin and Smith 1975), wherein some bacteria are not killed but are only inhibited from growing. This is well documented with nalidixic acid (Crumplin and Smith 1975) and also reported with other antibiotics such as β-lactam antibiotics (Ikeda and Nishino 1988: Ikeda *et al.* 1990).The importance of other technical factors affecting the efficiency of *in vitro* bactericidal tests was demonstrated by Taylor *et al.* (1983).

Recently, Singh *et al.* (2009) studied the dose-dependent killing of biofilms and planktonic cells of *Staphylococcus aureus* with five different antibiotics by treating them with each antibiotic at different concentrations for 48 h. They observed that the planktonic cells could be completely killed by oxacillin, cefotaxime, ciprofloxacin and vancomycin, leaving no persisters. With amikacin, the majority of the survivors were small colony variants that were highly resistant to the antibiotic. They also reported that *Staphylococcus aureus* biofilms could not be sterilized completely by any of these antibiotics. The absence of persisters in the planktonic cultures is in contrast with previous reports that the planktonic cells produce persisters that act as a nucleus for chronic infections. The reason for the absence of persisters in the above case can be attributed to the more ideal conditions for killing provided in the experiments i.e., dose-dependent killing for 48 h.

In vitro experiments demonstrating phenotypic shift have not considered most of the technical factors, nor have they provided the optimal conditions for 100% bacterial killing. Further, they have assumed that the bacteria that survived the antibiotic treatment have the capability to undergo a phenotypic shift. Researchers had incubated bacterial cultures using a single concentration of the antibiotic (Keren *et al.* 2004a; Balaban *et al.* 2004; Shah *et al.* 2006) against a specific inoculum size for a very short incubation time (Keren *et al.* 2004a; Lewis 2005; Balaban *et al.* 2004). Had the researchers incubated the bacterial culture with ampicillin for 24-48 h (at the same antibiotic concentration and inoculum size used in their experiments), they would have noticed one of three conditions.

1. Complete sterilization of the culture without any persisters;
2. Re-growth of the remaining survivors even in the presence of antibiotics;
3. Or, less frequently, the surviving bacteria remaining in the dormant stage itself without growth.

However, had these researchers tested different antibiotic concentrations and inoculum sizes followed by incubation for 24-48 h, they would have noticed the optimal conditions for complete sterilization. Thus, the perceived phenotypic shift demonstrated in many experiments may be the result of the failure of antibiotics to kill all bacteria owing to the suboptimal conditions of killing provided in the experiments.

3. Persisters may show increased MIC if incubated for a longer time.

It is reported that persisters exhibit the same MIC as the parent bacterial culture (Lewis 2007). Again, this may be true only for short incubation periods and at low inoculum sizes. At low inoculum sizes (10^5 or 10^6 bacteria/ml), persisters are either absent or very low in number. If some persisters are present, they may either remain dormant or die or regrow after 24-48 h. But at high inoculum sizes (10^8 bacteria/ml), the number of persisters can reach up to 10^5 to 10^6/ml (up to 1% of the total population). Those bacteria may not grow during the short incubation period of 3-6 h. However, if incubated for 24-48 h, they may repopulate the culture (depending on the type of antibiotic), thus changing the MIC. A lack of change of MIC in the above experiments is only due to the short incubation time during which the survivors were only inhibited from growth.

Hacek *et al.* (1999) reported a high level of reproducibility for the minimal bactericidal concentration (MBC) values. They found that 207 of 224 tests were reproducible at the 24 h subculture point for MBC assay. Of the remaining 17 tests, 16 were reproducible when subcultured for another 24 h. Thus, 223 of 224 tests were reproducible at the 48 h subculture point for MBC assay. This shows that the results obtained even after 24 h of incubation may not always be reproducible. A 3-6 h incubation period may not be sufficient to determine whether the persisters remain in a dormant stage in the presence of antibiotics or are capable of regrowth after a period of adaptation.

4. *In vitro* bacterial killing results cannot be extrapolated as such to *in vivo* conditions.

The experiments demonstrating phenotypic shift have used only a static approach; i.e., a culture of bacteria incubated with a single dose and a single concentration of antibiotic for a relatively small period of time. However, this static approach has a number of disadvantages (Mueller *et al.* 2004). *In vivo*, antibiotics are administered as multiple doses for several days. Antimicrobial efficacy results from the exposure of bacteria to variable antibiotic concentrations (Meuller *et al.* 2004). *In vitro* conditions with a single dose and concentration of antibiotics may not reflect the true dynamic situation in the target organ and hence the result can not be extrapolated as such to *in vivo* conditions.

As reviewed by Lewis (2007: 53), it may be possible to sterilize an infection by using a simple approach:

> *A disarmingly simple approach to sterilize an infection was first proposed by Bigger in 1944. The proposal is to kill bacterial cells with a high dose of an antibiotic, then allow the antibiotic concentration to decrease, which will enable persisters to resuscitate and start to grow. If a second dose of antibiotic is administered shortly after persisters start to grow, a complete sterilization might be achieved. This approach is successful in vitro, and a P. aeruginosa biofilm can essentially be sterilized with 2 consecutive applications of a fluoroquinolone (K. L., unpublished observations). Perhaps understandably, this approach has not been received with enthusiasm by specialists in clinical microbiology.*

Is this not the same 'approach' we follow during the treatment of bacterial infections with antibiotics? After the first dose of the antibiotic, the concentration of the antibiotic decreases depending on its half-life and is followed by the next dose. In clinical infections, most antibiotics are administered as multiple doses for several days. If a *P. aeruginosa* biofilm can essentially be sterilized with 2 consecutive applications of a fluoroquinolone or if a second dose of antibiotic can kill all persisters and achieve complete sterilization, what is the clinical significance of persisters?

5. Most bacterial infections are cured after antibiotic treatment. What are those 'ideal conditions' that cure infection without leaving any persisters thus preventing chronic infections? In the same manner, many chronic infections can be treated successfully without recurrence. What are those ideal conditions that kill all persisters in such chronic cases?
Lewis (2007: 54) continues:

> *The goal of established therapies is to maintain the plasma level of an antibiotic at a maximum concentration, in order to discourage the development of resistance. Most importantly, an optimal pulse-dosing regimen would probably vary from patient to patient. However, it seems that some patients might have inadvertently taken solving the problem of intractable persistent infections into their own hands. Individuals who suffer from persistent infections that require a lengthy therapy are often cured, but why a year-long regimen is better than a month-long one is unclear. An efficacious fluctuating dose of antibiotics administered serendipitously by the patient might be responsible for persister eradication in these cases. The patients might adjust drug dosing simply through being absent-minded, which sooner or later could produce the perfect drug-administration regimen. Curing persistent infections might therefore result from patient non-compliance. Analysing how persistent infections are cured might shed light on the likelihood of developing a rational regimen for the pulse-dosing sterilization of infection.*

There are two flaws in the above statement. First, maintaining the plasma level of antibiotics at maximum concentration is not always the goal of antibiotic therapy. This goal may be true only for those antibiotics exhibiting concentration-dependent killing. For time-dependent killing antibiotics, t>MIC is the important parameter. Second, one of the reasons for the development of drug resistance leading to antibiotic treatment failure is the non-adherence to antibiotic regimen by patients (Addington 1979; Kardas 2002; Zhao and Drlica 2001; Riley 1993; Anastasio *et al.* 1994). Hence it can invite criticism when suggested that serendipity or patient non-compliance can cure persistent infections in some cases. Can this suggestion be supported by any PK/PD models?

6. Continuous infusion of β-lactam antibiotics may offer more favorable PK/PD parameters over intermittent administration.

It is reported that the continuous infusion of β-lactam antibiotics offers more favorable PK/PD parameters over intermittent administration, especially against less susceptible organisms (Roberts *et al.* 2007; van Zanten *et al.* 2007; Lubasch *et al.* 2003; Lorente *et al.* 2009; Roberts *et al.* 2009; Hughes *et al.* 2009; Krueger *et al.* 2005; Jacqueline *et al.* 2002). In continuous infusion, the concentration of the antibiotic never reaches a trough level (Cmin), thus providing superior PK/PD parameters (Roberts *et al.* 2007). Persisters may never get a chance to resuscitate during the continuous infusion treatment period, whereas during intermittent administration, they may start to regrow when the drug concentration falls below a critical level, making them more susceptible to antibiotic action. Why is continuous infusion advantageous over intermittent administration, even though the persisters do not get a chance to resuscitate during the treatment period?

7. Phenotypic shift of persisters is related to the 'inoculum effect'.

The inoculum effect refers to the increase in the MIC when a higher than standard bacterial inoculum is used. The inoculum effect is due to density-dependent decline in antibiotic activity, which results from the ability of bacteria to produce an enzyme that can hydrolyze the antibiotic (see Craig *et al.* 2004 for a detailed review on this aspect of inoculum effect), decrease in the antibiotic concentration per cell (Udekwu *et al.* 2009), trapping of the antibiotic molecules in the cell wall (Sieradzki and Tomasz 2006; Cui *et al.* 2006) and the binding of antibiotics to the D-alanyl-D-alanine residues located in the cell wall (Sieradzki and Tomasz 2006). Concentration of some antibiotics in liquid medium may decrease gradually with high inoculum size (Yanagisawa *et al.* 2009, Sieradzki and Tomasz 2006; Cui *et al.* 2006) For example, the decline in vancomycin concentration for the first 8 h may result in bacteriostasis, whereas by 24 h, the concentration of antibiotic in the medium may reduce further, resulting in the regrowth of bacteria (Yanagisawa *et al.* 2009). Thus, at a high inoculum size, even though a large number of bacteria get killed initially, the remaining survivors are exposed to the subinhibitory concentration of antibiotics due to the decline in antibiotic activity. This concentration may inhibit the growth of bacteria, but may not be sufficient to kill them. However, after prolonged incubation, the survivors may start to grow again due to further reduction in antibiotic concentration and/or due to the ability of bacteria to adapt to the new condition. Exposure to subinhibitory concentrations of antibiotics results in a bacterial stress response, which will be discussed below.

Upon close examination, it can be noticed that the phenotypic shift of persisters is related to the inoculum effect. At low inoculum size, the number of persisters is low. As the inoculum size increases, the number of persisters also increases. They may remain dormant during the first few hours of incubation following exposure to subinhibitory concentrations of antibiotics, but may regrow on further incubation or on transfer to fresh medium without antibiotics. Thus, the re-growth of bacteria following transfer to fresh medium without antibiotics may not be due to the phenotypic shift of persisters, but rather can be due to the inoculum effect only. Since inoculum effect is mainly an *in vitro* phenomenon, the phenotypic shift of persisters may not have much clinical significance.

8. The number of persisters is determined by the type of antibiotics.

When the logarithmic phase of planktonic *Pseudomonas aeroginosa* was treated with carbenicillin at 600µg/ml for 6 h, 0.1% of bacteria survived whereas ofloxacin (15µg/ml) resulted in 0.0001% persisters. Treatment with tobramycin, on the other hand, completely killed all bacteria even at 4µg/ml (Spoering and Lewis 2001). If phenotypic shift is a property exhibited by a bacterial population, why are there large differences in the number of persisters with different antibiotics?

The difference in the number of persisters can be explained on the basis of the antibiotic property in exhibiting the inoculum effect (for the inoculum effect exhibited by different antibiotics, see Craig *et al.* 2004). Antibiotics that exhibit minimal inoculum effect such as aminoglycosides (e.g. tobramycin) may kill all exponentially growing bacteria. On the other hand, treatment with antibiotics that show a significant inoculum effect like ampicillin or carbenicillin may produce more survivors.

9. The number of persisters is not always small.

It has been reported that the number of persisters is very small (less than 1% of the bacterial population) (Lewis 2007). However, this is true only at certain inoculum sizes. At a very low inoculum size, there may not be any persisters. For a standard inoculum size, the number of persisters may be less than 1%. However, as the size increases, number of persisters also increases. The effect of inoculum size on the number of survivors was recently demonstrated by Udekwu et al. (2009), using six different antibiotics. In the case of ciprofloxacin, the number of survivors was less than 1% at an inoculum size of $2x10^6$ cfu/ml or below. But at an inoculum size of $9x10^6$ cfu/ml or above, the number of survivors was close to 10%. The same was true for gentamicin. But for oxacillin, the effect of the inoculum size was drastic. At $2x10^6$ cfu/ml or below, the number of persisters was close to 1%, but at an inoculum size of $9x10^6$ cfu/ml or above, it reached more than 20-30%. As stated above, this indicates that *in vitro* persisters are only the result of the inoculum effect and depend on the type of antibiotics. Because the number of persisters at an inoculum size of $9x10^6$ cfu/ml is more than 10% with all the three antibiotics, does it mean that these antibiotics are ineffective *in vivo*?

10. It is still safe to use bacteriostatic drugs for treatment of many bacterial infections even though they may leave more persisters than bactericidal drugs

Antibiotics, in general, can be divided into bactericidal and bacteriostatic drugs, depending on their mechanism of action. Bactericidal drugs like penicillin, aminoglycosides etc., act by killing the bacteria whereas bacteriostatic drugs like tetracycline, chloramphenicol, macrolides etc., act by inhibiting bacterial growth. However, it is difficult to classify antibiotics into true bactericidal or bacteriostatic antibiotics (Pankey and Sabath 2004). Bactericidal drugs that kill a particular organism or a strain of bacteria can be bacteriostatic on another organism (Boswell *et al.* 2002; Pankey and Sabath 2004), to the same organism at different conditions (Lleo *et al.* 1987) or at a different concentration of the antibiotic (Moellering *et al.* 1972). Similarly, bacteriostatic drugs also kill bacteria (Rahal and Simberkoff 1979; Zabransky *et al.* 1973) depending on the concentration, time of incubation and the sensitivity of the organism. Many bacteriostatic drugs kill 90-99% of the bacteria after 18-24 h, but do not kill 99.9% of them to be termed as bactericidal drugs (for a detailed review of the clinical significance of bactericidal and bacteriostatic antibiotics, see Pankey and Sabath 2004). Here again, the survivors remain in a dormant stage in the presence of the bacteriostatic antibiotic. Even though the static drugs, in general, do not kill 99.9% of bacteria, they are powerful antibiotics widely used in treatment of many bacterial infections.

If persisters are responsible for recurrent infections, one has to assume that the use of bacteriostatic drugs almost always carries the risk of chronic or recurrent infections. Since this is not the case, 100% *in vitro* bacterial killing may not be a requirement for the successful treatment of many bacterial infections. Bactericidal action may be necessary in some clinical situations like endocarditis, osteo-myelitis, meningitis, etc. (Pankey and Sabath 2004), but here again, complete sterilization by a single dose of antibiotics is not required.

11. 100% *in vitro* bacterial killing is not an absolute necessity for successful treatment of bacterial infections *in vivo*.

Persisters are implicated in chronic recurrent infections and in biofilms because the antibiotic treatment may leave a small subset of persisters which may later repopulate and cause infections (Levin and Rozen 2006; Lewis 2007; Dhar and McKinney 2007). This small subpopulation is considered to be the nucleus for chronic infections. However, the significance of these *in vitro* survivors is questionable as they may not affect the outcome of treatment.

If 100% killing is necessary, how can one justify the determina-tion of dosage regimen of an antibiotic using PK/PD parameters based on MIC? After all, MIC only indicates that bacteria are inhibit-ed from growth. The important PK/PD parameters such as the percentage of time above the MIC (t > MIC), ratio of peak concen-tration to MIC (Cmax/MIC) and the ratio of the area under the curve to MIC (AUC/MIC) are based on MIC. If 100% *in vitro* killing is a necessity, will it be more appropriate to use MBC values? Even MBC may not be ideal as, by definition, it kills only 99.9% of bacteria. Should we consider MBC_{100} as the single most important parameter that determine the *in vivo* antibiotic efficacy?

12. Most of the experiments demonstrating phenotypic shift of persisters are *in vitro* experiments.

The only currently available information on *in vivo* persisters is that of *Candida albicans* and *C. glabrata* isolated from cancer patients (LaFleur *et al.* 2010). Those patients were on long-term treatment with topical chlorhexidine as they were at high risk of oral candidiasis. Persister levels were higher in patients with long-term carriage than in patients with transient carriage. Fifteen *hip* mutants were isolated from different patients, indicating that *hip* mutants are involved in recalcitrant fungal infections.

However, these data do not support the hypothesis that a phenotypic shift of persisters is responsible for recalcitrant infections. *hip* mutants cannot be considered true persisters as they have an underlying genetic change, whereas persisters are non-mutants, exhibiting a phenotypic trait. Moreover, the strains isolated from those patients did not show any signs of infections, as admitted by the authors. Hence, it is not known whether the presence of these mutants can cause chronic infections.

13. A large set of diverse genes is implicated in persistence.

HipA is not the only protein implicated in persistence. Over-expression of toxins such as RelA, MazF, YgiU etc. also generate persisters. Genes implicated in persistence that are unrelated to TA systems include p*ho*U (Li and Zhang 2007), *glpD, plsB* (Spoering *et al.* 2006), *rpoS* (Murakami *et al.* 2005) *dnaJ, dnaK, surA, ygfA, yigB, apaH, fis, hns, hnr* and *dksA* (Hansen *et al.* 2008). Thus, specificity of HipA or other toxins in the generation of persisters is doubtful.

Vasquez-Laslop *et al.* (2006) reported that proteins unrelated to TA systems such as DnaJ or PmrC, when over-expressed, become toxic to cells and generate a high frequency of persisters. This raises the question whether the *in vitro* over-expression of proteins is a reliable method to explain the phenotypic shift of persisters.

14. Bacteria that survive antibiotics *in vitro* may not have mutations in *hipA* or *hipB*.

After plating approximately 10^{10} bacteria on agar plates containing different concentrations of ciprofloxacin for 96 h, Marcusson *et al.* (2005) found that some bacteria could survive the lethal action of the antibiotic. However, the survival of this subpopulation was not associated with any mutation in the *hipA* or *hipB* genes, indicating that the persistence of bacteria in the presence of antibiotics may not be due to *hipA* or *hipB* mutations.

15. Generation of some persisters depends on the 'paradoxical effect' of antibiotics.

An inducible mechanism of persister formation mediated by SOS response was recently reported (Dorr *et al.* 2009). It was found that, up to a critical concentration of 0.5 µg/ml, an increase in ciprofloxacin concentration increased SOS induction but reduced the number of persisters. However, further increase in ciprofloxacin concentration above 0.5 µg/ml increased both SOS induction and the number of persisters. It was suggested that the bacterial persistence in the presence of ciprofloxacin was dependent on a functional SOS response. However, since all cells exposed to ciprofloxacin could induce SOS, but only a fraction became persisters, it was suggested that a specific high or low level of SOS induction was required for persister formation (Dorr *et al.* 2009).

It is interesting to note that the above phenomenon is the 'paradoxical effect' of antibiotics, which was reported much earlier (Crumplin and Smith 1975; Piddock *et al.* 1990). A paradoxical effect (reduction in bactericidal activity with high antibiotic concentrations of antibiotics above a critical level) is well documented with nalidixic acid (Crumplin and Smith 1975) and other quinolones (Piddock *et al.* 1990; Udekwu *et al.* 2009). The *in vivo* significance of the paradoxical effect is not clear, even though a few reports have suggested that it can occur *in vivo* also (Voorn *et al.* 1994; Griffiths and Green 1985; Ikeda *et al.* 1990). The paradoxical effect of quinolones is attributed to reduced RNA or protein synthesis at high antibiotic concentration (Crumplin and Smith 1975). Recently, it was proposed that the paradoxical effect depends on Lon protease activity and that neither protein synthesis nor SOS response is required for this phenomenon (Malik *et al.* 2009).

Even though quinolones induce an SOS response (Phillips *et al.* 1987; Piddock *et al.* 1990; Lewin *et al.* 1989; Dorr *et al.* 2009), the SOS-DNA repair system may not play any role in protecting the bacteria from damage or cause the death of bacteria (Lewin *et al.* 1989; Malik *et al.* 2009). Piddock *et al.* (1990) reported that the primary mechanism of quinolone action is independent of the SOS response. More importantly, they noticed that, even though the total survivors were more at higher concentrations, prolonged exposure of bacteria at high concentrations reduced the number of viable bacteria.

Thus, Dorr *et al.* (2009) failed to correlate the paradoxical effect and inducible persister formation by ciprofloxacin. There is an ideal concentration of quinolone where survivors are minimal. Concentrations above or below this critical level may generate more survivors *in vitro* and induce an SOS response. However, the SOS response may not be responsible for the formation of these persisters. Similarly, the paradoxical effect may not have much clinical significance *in vivo* when the treatment is prolonged for days.

16. Persisters of exponential and stationary phase cultures, biofilms and *hipA* mutants are different.

High inoculum exponential phase cultures, stationary phase cultures and biofilms may generate many persisters after antibiotic treatment. Similarly, a high frequency of persisters is produced by *hipA* mutants. Even though there is a tendency to explain these persisters with a common phenomenon, they are generated by different mechanisms. Exponential phase persisters can be the result of an inoculum effect only. Stationary phase antibiotic tolerance can be attributed to starvation or to a general stress response (Fung *et al.* 2010), a slow growth rate and the reduced expression of some proteins (Stevens *et al.* 1993). Biofilm resistance closely resembles stationary phase resistance and can be due to reduced antibiotic penetration, decreased growth rate and nutrient starvation (Costerton *et al.* 1999; Anderson and O'Toole 2008; Anderl *et al.* 2003). Even though *hipA* mutants generate many survivors, they cannot be considered true persisters since persisters are non-mutants that exhibit a phenotypic trait.

17. The relationship between *hipA* and the phenotypic shift of persisters is not clear.

Overexpression of HipA produces a high frequency of persisters (Falla and Chopra 1998; Keren *et al.* 2004a). Similarly, mutation in *hipA* also produces a high frequency of persisters (Moyed and Bertrand 1983). Based on these findings, it was proposed that the stochastic fluctuations in the expression of HipA could be responsible for the generation of persisters (Keren *et al.* 2004b; Lewis 2007; Lewis 2008). However, researchers have not provided any evidence to demonstrate such stochastic fluctuations in the levels of HipA following antibiotic treatment. Moreover, overexpression of a number of proteins can generate a high frequency of persisters, as indicated above. Thus the role of *hipA* in the phenotypic shift of persisters is vague and obscure.

18. Persistence, as demonstrated in many experiments, is a bacterial stress response rather than a phenotypic trait.

Bacteria exposed to unfavorable conditions elicit a stress response that includes inhibition of bacterial growth, cellular damage repair and expression of a different set of proteins that may help overcome the stressful conditions. During this period of stress response, bacteria may become tolerant, at least to some antibiotics. For example, the stationary phase induces a starvation response, which makes bacteria tolerant to many antibiotics (Fung *et al.* 2010). Quinolones cause double strand breaks in DNA, which may induce a bacterial SOS response (Dorr *et al.* 2009). Similarly, overexpression of a number of proteins like MarA (Asako *et al.* 1997), SoxS (Nakajima *et al.* 1995b), or RobA (Nakajima *et al.* 1995a), in response to unfavorable conditions, results in antibiotic tolerance.

Persistence, as demonstrated in many experiments, is a general stress response rather than a phenotypic trait and is not limited to antibiotic exposure; rather it is seen with a variety of different stressors. For example, treatment of a bacterial culture with hemin produces some survivors that do not grow in the presence of hemin but start to regrow once the hemin is removed (Nitzan *et al.* 1987). In the presence of stressors, some bacteria may remain dormant transiently but may either regrow after a period of adaptation or undergo death in the continued presence of the stressor at optimal concentrations. The number of bacteria that survive depend on the inoculum size, the concentration of stressors and the total time of exposure. Thus, phenotypic shift, as demonstrated by *in vitro* experiments, can be observed with a wide range of stressors (i.e., not only with antibiotics). Bacteria that enter the body encounter a number of stress conditions such as nutrient limitation, change in pH, change in oxygen tension etc. and persisters can be produced in such conditions also. Are those persisters also responsible for chronic infections? If different stressors are capable of producing persisters, why is persistence important only in antibiotic therapy?

Recently, a mechanism of persistence was suggested based on the structure of HipA and HipA-HipB-DNA complexes (Schumacher *et al.* 2009). As per the model, phosphorylation of the translation factor EF-Tu by HipA blocks aminoacyl-tRNA binding by EF-Tu, which leads to cell stasis and persistence. It was suggested that the inhibitors that target binding sites of HipA could be useful against persistence (Schumacher *et al.* 2009). However, the specific role of HipA in antibiotic tolerance is not clear. EF-Tu can also function as a chaperone and protect bacteria from many types of stress including heat shock stress (Caldas *et al.* 1998). Similarly, the *hipBA* system may also be involved in the survival of bacterial cells in the stationary phase (Kawano *et al.* 2009), indicating that it may play a role in general stress responses.

Conclusion: Phenotypic shift of persisters is an *in vitro* illusion created artificially through retrospective thinking.

It is known that biofilms are difficult to treat with conventional antibiotic therapy. It has also been reported that biofilms are responsible for 65% of bacterial infections in developed countries (Costerton 2001). Even though biofilms are tolerant to antibiotics, planktonic cells separated from the biofilms may remain sensitive. Using this knowledge, an *in vitro* illusion has been created to demonstrate that recurrent infections and those infections by biofilms are caused by a phenotypic shift of persisters. A general bacterial stress response that may not have much clinical significance is made into a topic of utmost public health importance through this illusion.

Phenotypic shift of persisters can be an artificial phenomenon and a theoretical concept demonstrated only in *in vitro* conditions. Experimental methods are questionable, and it is not clear whether the regrowth of bacteria after antibiotic removal is indeed due to a small subset of the population undergoing a phenotypic shift or simply the result of the failure of antibiotics to kill all bacteria due to suboptimal conditions of killing provided in the experiments. It is also not clear whether a subpopulation of less than 1% persisters is responsible for chronic or recurrent infections since a 100% bacterial killing by antibiotics *in vitro* may not be a requirement for successful treatment of bacterial infections. The specific role of proteins such as HipA, proposed to be involved in the formation of persisters, is also questionable, as over-expression of other proteins unrelated to persister formation can also give rise to a high frequency of persisters *in vitro*. Lack of data on *in vivo* persisters in chronic infections reinforces this suggestion.

Experimental findings indicate that persistence (as demonstrated in experiments) can be a general stress response only. At a low inoculum size (10^5-10^6/ml), bacteria are exposed to antibiotics uniformly, resulting in their death without leaving any survivors. As the inoculum size increases, the majority of them are killed, whereas the remaining survivors are exposed to a subinhibitory antibiotic concentration due to density-dependent antibiotic decay. The bacteria may remain dormant initially and elicit a general bacterial stress response to overcome the unfavorable environment. They may start to regrow after a period of adaptation (thus changing the MIC) or if they are transferred to fresh medium without antibiotics. For quinolones, the same is true up to a critical antibiotic concentration. Above this concentration (the supra-inhibitory concentration), the number of survivors increases due to the paradoxical effect. Here also, a bacterial stress response (SOS response) will be elicited, but the SOS response may not have any role in persister formation. Thus, both the inoculum effect and the paradoxical effect may generate many survivors *in vitro*, but they may not have much significance *in vivo*.

Even though I question the current knowledge on persisters, I also acknowledge the existence of 'true persisters', which are slow-dividing or non-dividing bacteria present in the culture at very low numbers. They are tolerant to antibiotics and are not mutants. However, they may not revert to normal growth after the removal of antibiotics; i.e., they do not undergo phenotypic shift and hence may not have much clinical significance. I hypothesize that these slow-dividing bacteria are senescent bacteria, which will be discussed later.

References

Addington, W. W. (1979). Patient compliance: the most serious remaining problem in the control of tuberculosis in the United States. *Chest* 76(6 Suppl), 741-3.

Al-Dhaheri, R. S., and Douglas, L. J. (2008). Absence of amphotericin B-tolerant persister cells in biofilms of some Candida species. *Antimicrob Agents Chemother* 52(5), 1884-7.

Al-Dhaheri, R. S., and Douglas, L. J. (2010). Apoptosis in Candida biofilms exposed to amphotericin B. *J Med Microbiol* 59(Pt 2), 149-57.

Amitai, S., Kolodkin-Gal, I., Hananya-Meltabashi, M., Sacher, A., and Engelberg-Kulka, H. (2009). Escherichia coli MazF leads to the simultaneous selective synthesis of both "death proteins" and "survival proteins". *PLoS Genet* 5(3), e1000390.

Amitai, S., Yassin, Y., and Engelberg-Kulka, H. (2004). MazF-mediated cell death in Escherichia coli: a point of no return. *J Bacteriol* 186(24), 8295-300.

Anastasio, G. D., Little, J. M., Jr., Robinson, M. D., Pettice, Y. L., Leitch, B. B., and Norton, H. J. (1994). Impact of compliance and side effects on the clinical outcome of patients treated with oral erythromycin. *Pharmacotherapy* 14(2), 229-34.

Anderl, J. N., Zahller, J., Roe, F., and Stewart, P. S. (2003). Role of nutrient limitation and stationary-phase existence in Klebsiella pneumoniae biofilm resistance to ampicillin and ciprofloxacin. *Antimicrob Agents Chemother* 47(4), 1251-6.

Anderson, G. G., and O'Toole, G. A. (2008). Innate and induced resistance mechanisms of bacterial biofilms. *Curr Top Microbiol Immunol* 322, 85-105.

Asako, H., Nakajima, H., Kobayashi, K., Kobayashi, M., and Aono, R. (1997). Organic solvent tolerance and antibiotic resistance increased by overexpression of marA in Escherichia coli. *Appl Environ Microbiol* 63(4), 1428-33.

Balaban, N. Q., Merrin, J., Chait, R., Kowalik, L., and Leibler, S. (2004). Bacterial persistence as a phenotypic switch. *Science* 305(5690), 1622-5.

Bech, F. W., Jorgensen, S. T., Diderichsen, B., and Karlstrom, O. H. (1985). Sequence of the relB transcription unit from Escherichia coli and identification of the relB gene. *EMBO J* 4(4), 1059-66.

Bigger, J. W. (1944). Treatment of staphylococcal infections with penicillin by intermittent sterilization. *Lancet* ii, 497-500.

Black, D. S., Irwin, B., and Moyed, H. S. (1994). Autoregulation of hip, an operon that affects lethality due to inhibition of peptidoglycan or DNA synthesis. *J Bacteriol* 176(13), 4081-91.

Boswell, F. J., Ashby, J. P., Andrews, J. M., and Wise, R. (2002). Effect of protein binding on the *in vitro* activity and pharmacodynamics of faropenem. *J Antimicrob Chemother* 50(4), 525-32.

Brook, I. (1989). Inoculum effect. *Rev Infect Dis* 11(3), 361-8.

Bulger, R. R., and Washington, J. A., 2nd (1980). Effect of inoculum size and beta-lactamase production on *in vitro* activity of new cephalosporins against Haemophilus species. *Antimicrob Agents Chemother* 17(3), 393-6.

Caldas, T. D., El Yaagoubi, A., and Richarme, G. (1998). Chaperone properties of bacterial elongation factor EF-Tu. *J Biol Chem* 273(19), 11478-82.

Correia, F. F., D'Onofrio, A., Rejtar, T., Li, L., Karger, B. L., Makarova, K., Koonin, E. V., and Lewis, K. (2006). Kinase activity of overexpressed HipA is required for growth arrest and multidrug tolerance in Escherichia coli. *J Bacteriol* 188(24), 8360-7.

Costerton, J. W. (2001). Cystic fibrosis pathogenesis and the role of biofilms in persistent infection. *Trends Microbiol* 9(2), 50-2.

Costerton, J. W., Stewart, P. S., and Greenberg, E. P. (1999). Bacterial biofilms: a common cause of persistent infections. *Science* 284(5418), 1318-22.

Craig, W. A., Bhavnani, S. M., and Ambrose, P. G. (2004). The inoculum effect: fact or artifact? *Diagn Microbiol Infect Dis* 50(4), 229-30.

Crumplin, G. C., and Smith, J. T. (1975). Nalidixic acid: an antibacterial paradox. *Antimicrob Agents Chemother* 8(3), 251-61.

Cui, L., Iwamoto, A., Lian, J. Q., Neoh, H. M., Maruyama, T., Horikawa, Y., and Hiramatsu, K. (2006). Novel mechanism of antibiotic resistance originating in vancomycin-intermediate Staphylococcus aureus. *Antimicrob Agents Chemother* 50(2), 428-38.

Darouiche, R. O., Dhir, A., Miller, A. J., Landon, G. C., Raad, II, and Musher, D. M. (1994). Vancomycin penetration into biofilm covering infected prostheses and effect on bacteria. *J Infect Dis* 170(3), 720-3.

Davey, P. G., and Barza, M. (1987). The inoculum effect with gram-negative bacteria *in vitro* and *in vivo*. *J Antimicrob Chemother* 20(5), 639-44.

Debbia, E. A., Roveta, S., Schito, A. M., Gualco, L., and Marchese, A. (2001). Antibiotic persistence: the role of spontaneous DNA repair response. *Microb Drug Resist* 7(4), 335-42.

del Pozo, J. L., and Patel, R. (2007). The challenge of treating biofilm-associated bacterial infections. *Clin Pharmacol Ther* 82(2), 204-9.

Dhar, N., and McKinney, J. D. (2007). Microbial phenotypic heterogeneity and antibiotic tolerance. *Curr Opin Microbiol* 10(1), 30-8.

Dorr, T., Lewis, K., and Vulic, M. (2009). SOS response induces persistence to fluoroquinolones in Escherichia coli. *PLoS Genet* 5(12), e1000760.

Eng, R. H., Cherubin, C., Smith, S. M., and Buccini, F. (1985). Inoculum effect of beta-lactam antibiotics on Enterobacteriaceae. *Antimicrob Agents Chemother* 28(5), 601-6.

Engelberg-Kulka, H., Hazan, R., and Amitai, S. (2005). mazEF: a chromosomal toxin-antitoxin module that triggers programmed cell death in bacteria. *J Cell Sci* 118(Pt 19), 4327-32.

Falla, T. J., and Chopra, I. (1998). Joint tolerance to beta-lactam and fluoroquinolone antibiotics in Escherichia coli results from overexpression of hipA. *Antimicrob Agents Chemother* 42(12), 3282-4.

Firsov, A. A., Ruble, M., Gilbert, D., Saverino, D., Manzano, B., Medeiros, A. A., and Zinner, S. H. (1997). Net effect of inoculum size on antimicrobial action of ampicillin-sulbactam: studies using an *in vitro* dynamic model. *Antimicrob Agents Chemother* 41(1), 7-12.

Fung, D. K., Chan, E. W., Chin, M. L., and Chan, R. C. (2010). Delineation of a bacterial starvation stress response network which can mediate antibiotic tolerance development. *Antimicrob Agents Chemother* 54(3), 1082-93.

Gefen, O., Gabay, C., Mumcuoglu, M., Engel, G., and Balaban, N. Q. (2008). Single-cell protein induction dynamics reveals a period of vulnerability to antibiotics in persister bacteria. *Proc Natl Acad Sci U S A* 105(16), 6145-9.

Gotfredsen, M., and Gerdes, K. (1998). The Escherichia coli relBE genes belong to a new toxin-antitoxin gene family. *Mol Microbiol* 29(4), 1065-76.

Griffiths, L. R., and Green, H. T. (1985). Paradoxical effect of penicillin in-vivo. *J Antimicrob Chemother* 15(4), 507-8.

Hacek, D. M., Dressel, D. C., and Peterson, L. R. (1999). Highly reproducible bactericidal activity test results by using a modified National Committee for Clinical Laboratory Standards broth macrodilution technique. *J Clin Microbiol* 37(6), 1881-4.

Hansen, S., Lewis, K., and Vulic, M. (2008). Role of global regulators and nucleotide metabolism in antibiotic tolerance in Escherichia coli. *Antimicrob Agents Chemother* 52(8), 2718-26.

Harrison, J. J., Ceri, H., Roper, N. J., Badry, E. A., Sproule, K. M., and Turner, R. J. (2005). Persister cells mediate tolerance to metal oxyanions in Escherichia coli. *Microbiology* 151(Pt 10), 3181-95.

Hazan, R., Sat, B., and Engelberg-Kulka, H. (2004). Escherichia coli mazEF-mediated cell death is triggered by various stressful conditions. *J Bacteriol* 186(11), 3663-9.

Holm, S. E., Tornqvist, I. O., and Cars, O. (1990). Paradoxical effects of antibiotics. *Scand J Infect Dis Suppl* 74, 113-7.

Honda, R., Lowe, E. D., Dubinina, E., Skamnaki, V., Cook, A., Brown, N. R., and Johnson, L. N. (2005). The structure of cyclin E1/CDK2: implications for CDK2 activation and CDK2-independent roles. *EMBO J* 24(3), 452-63.

Hughes, D. W., Frei, C. R., Maxwell, P. R., Green, K., Patterson, J. E., Crawford, G. E., and Lewis, J. S., 2nd (2009). Continuous versus intermittent infusion of oxacillin for treatment of infective endocarditis caused by methicillin-susceptible Staphylococcus aureus. *Antimicrob Agents Chemother* 53(5), 2014-9.

Ikeda, Y., Fukuoka, Y., Motomura, K., Yasuda, T., and Nishino, T. (1990). Paradoxical activity of beta-lactam antibiotics against Proteus vulgaris in experimental infection in mice. *Antimicrob Agents Chemother* 34(1), 94-7.

Ikeda, Y., and Nishino, T. (1988). Paradoxical antibacterial activities of beta-lactams against Proteus vulgaris: mechanism of the paradoxical effect. *Antimicrob Agents Chemother* 32(7), 1073-7.

Jacqueline, C., Batard, E., Perez, L., Boutoille, D., Hamel, A., Caillon, J., Kergueris, M. F., Potel, G., and Bugnon, D. (2002). *In vivo* efficacy of continuous infusion versus intermittent dosing of linezolid compared to vancomycin in a methicillin-resistant Staphylococcus aureus rabbit endocarditis model. *Antimicrob Agents Chemother* 46(12), 3706-11.

Kaldalu, N., Mei, R., and Lewis, K. (2004). Killing by ampicillin and ofloxacin induces overlapping changes in Escherichia coli transcription profile. *Antimicrob Agents Chemother* 48(3), 890-6.

Kardas, P. (2002). Patient compliance with antibiotic treatment for respiratory tract infections. *J Antimicrob Chemother* 49(6), 897-903.

Kawano, H., Hirokawa, Y., and Mori, H. (2009). Long-term survival of Escherichia coli lacking the HipBA toxin-antitoxin system during prolonged cultivation. *Biosci Biotechnol Biochem* 73(1), 117-23.

Keren, I., Kaldalu, N., Spoering, A., Wang, Y., and Lewis, K. (2004a). Persister cells and tolerance to antimicrobials. *FEMS Microbiol Lett* 230(1), 13-8.

Keren, I., Shah, D., Spoering, A., Kaldalu, N., and Lewis, K. (2004b). Specialized persister cells and the mechanism of multidrug tolerance in Escherichia coli. *J Bacteriol* 186(24), 8172-80.

Kolodkin-Gal, I., and Engelberg-Kulka, H. (2009). The stationary-phase sigma factor sigma(S) is responsible for the resistance of Escherichia coli stationary-phase cells to mazEF-mediated cell death. *J Bacteriol* 191(9), 3177-82.

Kolodkin-Gal, I., Verdiger, R., Shlosberg-Fedida, A., and Engelberg-Kulka, H. (2009). A differential effect of E. coli toxin-antitoxin systems on cell death in liquid media and biofilm formation. *PLoS One* 4(8), e6785.

Konig, C., Simmen, H. P., and Blaser, J. (1998). Bacterial concentrations in pus and infected peritoneal fluid--implications for bactericidal activity of antibiotics. *J Antimicrob Chemother* 42(2), 227-32.

Korch, S. B., Henderson, T. A., and Hill, T. M. (2003). Characterization of the hipA7 allele of Escherichia coli and evidence that high persistence is governed by (p)ppGpp synthesis. *Mol Microbiol* 50(4), 1199-213.

Korch, S. B., and Hill, T. M. (2006). Ectopic overexpression of wild-type and mutant hipA genes in Escherichia coli: effects on macromolecular synthesis and persister formation. *J Bacteriol* 188(11), 3826-36.

Krueger, W. A., Bulitta, J., Kinzig-Schippers, M., Landersdorfer, C., Holzgrabe, U., Naber, K. G., Drusano, G. L., and Sorgel, F. (2005). Evaluation by monte carlo simulation of the pharmacokinetics of two doses of meropenem administered intermittently or as a continuous infusion in healthy volunteers. *Antimicrob Agents Chemother* 49(5), 1881-9.

Kussell, E., Kishony, R., Balaban, N. Q., and Leibler, S. (2005). Bacterial persistence: a model of survival in changing environments. *Genetics* 169(4), 1807-14.

LaFleur, M. D., Kumamoto, C. A., and Lewis, K. (2006). Candida albicans biofilms produce antifungal-tolerant persister cells. *Antimicrob Agents Chemother* 50(11), 3839-46.

Lafleur, M. D., Qi, Q., and Lewis, K. (2010). Patients with long-term oral carriage harbor high-persister mutants of Candida albicans. *Antimicrob Agents Chemother* 54(1), 39-44.

LaPlante, K. L., and Rybak, M. J. (2004). Impact of high-inoculum Staphylococcus aureus on the activities of nafcillin, vancomycin, linezolid, and daptomycin, alone and in combination with gentamicin, in an *in vitro* pharmacodynamic model. *Antimicrob Agents Chemother* 48(12), 4665-72.

Levin, B. R., and Rozen, D. E. (2006). Non-inherited antibiotic resistance. *Nat Rev Microbiol* 4(7), 556-62.

Lewin, C. S., Howard, B. M., Ratcliffe, N. T., and Smith, J. T. (1989). 4-quinolones and the SOS response. *J Med Microbiol* 29(2), 139-44.

Lewis, K. (2001). Riddle of biofilm resistance. *Antimicrob Agents Chemother* 45(4), 999-1007.

Lewis, K. (2005). Persister cells and the riddle of biofilm survival. *Biochemistry (Mosc)* 70(2), 267-74.

Lewis, K. (2007). Persister cells, dormancy and infectious disease. *Nat Rev Microbiol* 5(1), 48-56.

Lewis, K. (2008). Multidrug tolerance of biofilms and persister cells. *Curr Top Microbiol Immunol* 322, 107-31.

Li, Y., and Zhang, Y. (2007). PhoU is a persistence switch involved in persister formation and tolerance to multiple antibiotics and stresses in Escherichia coli. *Antimicrob Agents Chemother* 51(6), 2092-9.

Lleo, M. M., Canepari, P., Cornaglia, G., Fontana, R., and Satta, G. (1987). Bacteriostatic and bactericidal activities of beta-lactams against Streptococcus (Enterococcus) faecium are associated with saturation of different penicillin-binding proteins. *Antimicrob Agents Chemother* 31(10), 1618-26.

Lorente, L., Jimenez, A., Martin, M. M., Iribarren, J. L., Jimenez, J. J., and Mora, M. L. (2009). Clinical cure of ventilator-associated pneumonia treated with piperacillin/tazobactam administered by continuous or intermittent infusion. *Int J Antimicrob Agents* 33(5), 464-8.

Lubasch, A., Luck, S., Lode, H., Mauch, H., Lorenz, J., Bolcskei, P., and Welte, T. (2003). Optimizing ceftazidime pharmacodynamics in patients with acute exacerbation of severe chronic bronchitis. *J Antimicrob Chemother* 51(3), 659-64.

Mah, T. F., and O'Toole, G. A. (2001). Mechanisms of biofilm resistance to antimicrobial agents. *Trends Microbiol* 9(1), 34-9.

Malik, M., Capecci, J., and Drlica, K. (2009). Lon protease is essential for paradoxical survival of Escherichia coli exposed to high concentrations of quinolone. *Antimicrob Agents Chemother* 53(7), 3103-5.

Marcusson, L. L., Olofsson, S. K., Komp Lindgren, P., Cars, O., and Hughes, D. (2005). Mutant prevention concentrations of ciprofloxacin for urinary tract infection isolates of Escherichia coli. *J Antimicrob Chemother* 55(6), 938-43.

Masuda, G., Yajima, T., Nakamura, K., Yanagishita, T., and Hayashi, H. (1979). Comparative bactericidal activities of beta-lactam antibiotics determined in agar and broth media. *J Antibiot (Tokyo)* 32(11), 1168-73.

Metzger, S., Dror, I. B., Aizenman, E., Schreiber, G., Toone, M., Friesen, J. D., Cashel, M., and Glaser, G. (1988). The nucleotide sequence and characterization of the relA gene of Escherichia coli. *J Biol Chem* 263(30), 15699-704.

Mittenhuber, G. (1999). Occurrence of mazEF-like antitoxin/toxin systems in bacteria. *J Mol Microbiol Biotechnol* 1(2), 295-302.

Moellering, R. C., Jr., Medoff, G., Leech, I., Wennersten, C., and Kunz, L. J. (1972). Antibiotic synergism against Listeria monocytogenes. *Antimicrob Agents Chemother* 1(1), 30-4.

Moyed, H. S., and Bertrand, K. P. (1983). hipA, a newly recognized gene of Escherichia coli K-12 that affects frequency of persistence after inhibition of murein synthesis. *J Bacteriol* 155(2), 768-75.

Moyed, H. S., and Broderick, S. H. (1986). Molecular cloning and expression of hipA, a gene of Escherichia coli K-12 that affects frequency of persistence after inhibition of murein synthesis. *J Bacteriol* 166(2), 399-403.

Mueller, M., de la Pena, A., and Derendorf, H. (2004). Issues in pharmacokinetics and pharmacodynamics of anti-infective agents: kill curves versus MIC. *Antimicrob Agents Chemother* 48(2), 369-77.

Murakami, K., Ono, T., Viducic, D., Kayama, S., Mori, M., Hirota, K., Nemoto, K., and Miyake, Y. (2005). Role for rpoS gene of Pseudomonas aeruginosa in antibiotic tolerance. *FEMS Microbiol Lett* 242(1), 161-7.

Nakajima, H., Kobayashi, K., Kobayashi, M., Asako, H., and Aono, R. (1995a). Overexpression of the robA gene increases organic solvent tolerance and multiple antibiotic and heavy metal ion resistance in Escherichia coli. *Appl Environ Microbiol* 61(6), 2302-7.

Nakajima, H., Kobayashi, M., Negishi, T., and Aono, R. (1995b). soxRS gene increased the level of organic solvent tolerance in Escherichia coli. *Biosci Biotechnol Biochem* 59(7), 1323-5.

Nickel, J. C., Ruseska, I., Wright, J. B., and Costerton, J. W. (1985). Tobramycin resistance of Pseudomonas aeruginosa cells growing as a biofilm on urinary catheter material. *Antimicrob Agents Chemother* 27(4), 619-24.

Nitzan, Y., Ladan, H., and Malik, Z. (1987). Growth-Inhibitory Effect of Hemin on Staphylococci. *Current Microbiology* 14(5), 279-284.

Pandey, D. P., and Gerdes, K. (2005). Toxin-antitoxin loci are highly abundant in free-living but lost from host-associated prokaryotes. *Nucleic Acids Res* 33(3), 966-76.

Pankey, G. A., and Sabath, L. D. (2004). Clinical relevance of bacteriostatic versus bactericidal mechanisms of action in the treatment of Gram-positive bacterial infections. *Clin Infect Dis* 38(6), 864-70.

Patel, R. (2005). Biofilms and antimicrobial resistance. *Clin Orthop Relat Res*(437), 41-7.

Pedersen, K., Christensen, S. K., and Gerdes, K. (2002). Rapid induction and reversal of a bacteriostatic condition by controlled expression of toxins and antitoxins. *Mol Microbiol* 45(2), 501-10.

Peterson, L. R., and Shanholtzer, C. J. (1992). Tests for bactericidal effects of antimicrobial agents: technical performance and clinical relevance. *Clin Microbiol Rev* 5(4), 420-32.

Phillips, I., Culebras, E., Moreno, F., and Baquero, F. (1987). Induction of the SOS response by new 4-quinolones. *J Antimicrob Chemother* 20(5), 631-8.

Piddock, L. J., Walters, R. N., and Diver, J. M. (1990). Correlation of quinolone MIC and inhibition of DNA, RNA, and protein synthesis and induction of the SOS response in Escherichia coli. *Antimicrob Agents Chemother* 34(12), 2331-6.

Rahal, J. J., Jr., and Simberkoff, M. S. (1979). Bactericidal and bacteriostatic action of chloramphenicol against meningeal pathogens. *Antimicrob Agents Chemother* 16(1), 13-18.

Riley, L. W. (1993). Drug-resistant tuberculosis. *Clin Infect Dis* 17 Suppl 2, S442-6.

Roberts, J. A., Kirkpatrick, C. M., Roberts, M. S., Robertson, T. A., Dalley, A. J., and Lipman, J. (2009). Meropenem dosing in critically ill patients with sepsis and without renal dysfunction: intermittent bolus versus continuous administration? Monte Carlo dosing simulations and subcutaneous tissue distribution. *J Antimicrob Chemother* 64(1), 142-50.

Roberts, J. A., Paratz, J., Paratz, E., Krueger, W. A., and Lipman, J. (2007). Continuous infusion of beta-lactam antibiotics in severe infections: a review of its role. *Int J Antimicrob Agents* 30(1), 11-8.

Roostalu, J., Joers, A., Luidalepp, H., Kaldalu, N., and Tenson, T. (2008). Cell division in Escherichia coli cultures monitored at single cell resolution. *BMC Microbiol* 8, 68.

Sat, B., Hazan, R., Fisher, T., Khaner, H., Glaser, G., and Engelberg-Kulka, H. (2001). Programmed cell death in Escherichia coli: some antibiotics can trigger mazEF lethality. *J Bacteriol* 183(6), 2041-5.

Schembri, M. A., Kjaergaard, K., and Klemm, P. (2003). Global gene expression in Escherichia coli biofilms. *Mol Microbiol* 48(1), 253-67.

Schumacher, M. A., Piro, K. M., Xu, W., Hansen, S., Lewis, K., and Brennan, R. G. (2009). Molecular mechanisms of HipA-mediated multidrug tolerance and its neutralization by HipB. *Science* 323(5912), 396-401.

Shah, D., Zhang, Z., Khodursky, A., Kaldalu, N., Kurg, K., and Lewis, K. (2006). Persisters: a distinct physiological state of E. coli. *BMC Microbiol* 6, 53.

Sieradzki, K., and Tomasz, A. (2006). Inhibition of the autolytic system by vancomycin causes mimicry of vancomycin-intermediate Staphylococcus aureus-type resistance, cell concentration dependence of the MIC, and antibiotic tolerance in vancomycin-susceptible S. aureus. *Antimicrob Agents Chemother* 50(2), 527-33.

Singh, R., Ray, P., Das, A., and Sharma, M. (2009). Role of persisters and small-colony variants in antibiotic resistance of planktonic and biofilm-associated Staphylococcus aureus: an *in vitro* study. *J Med Microbiol* 58(Pt 8), 1067-73.

Spoering, A. L., and Lewis, K. (2001). Biofilms and planktonic cells of Pseudomonas aeruginosa have similar resistance to killing by antimicrobials. *J Bacteriol* 183(23), 6746-51.

Spoering, A. L., Vulic, M., and Lewis, K. (2006). GlpD and PlsB participate in persister cell formation in Escherichia coli. *J Bacteriol* 188(14), 5136-44.

Stevens, D. L., Yan, S., and Bryant, A. E. (1993). Penicillin-binding protein expression at different growth stages determines penicillin efficacy *in vitro* and *in vivo*: an explanation for the inoculum effect. *J Infect Dis* 167(6), 1401-5.

Stewart, P. S., and Franklin, M. J. (2008). Physiological heterogeneity in biofilms. *Nat Rev Microbiol* 6(3), 199-210.

Stone, G., Wood, P., Dixon, L., Keyhan, M., and Matin, A. (2002). Tetracycline rapidly reaches all the constituent cells of uropathogenic Escherichia coli biofilms. *Antimicrob Agents Chemother* 46(8), 2458-61.

Taylor, P. C., Schoenknecht, F. D., Sherris, J. C., and Linner, E. C. (1983). Determination of minimum bactericidal concentrations of oxacillin for Staphylococcus aureus: influence and significance of technical factors. *Antimicrob Agents Chemother* 23(1), 142-50.

Tsilibaris, V., Maenhaut-Michel, G., Mine, N., and Van Melderen, L. (2007). What is the benefit to Escherichia coli of having multiple toxin-antitoxin systems in its genome? *J Bacteriol* 189(17), 6101-8.

Udekwu, K. I., Parrish, N., Ankomah, P., Baquero, F., and Levin, B. R. (2009). Functional relationship between bacterial cell density and the efficacy of antibiotics. *J Antimicrob Chemother* 63(4), 745-57.

van Zanten, A. R., Oudijk, M., Nohlmans-Paulssen, M. K., van der Meer, Y. G., Girbes, A. R., and Polderman, K. H. (2007). Continuous vs. intermittent cefotaxime administration in patients with chronic obstructive pulmonary disease and respiratory tract infections: pharmacokinetics/pharmacodynamics, bacterial susceptibility and clinical efficacy. *Br J Clin Pharmacol* 63(1), 100-9.

Vazquez-Laslop, N., Lee, H., and Neyfakh, A. A. (2006). Increased persistence in Escherichia coli caused by controlled expression of toxins or other unrelated proteins. *J Bacteriol* 188(10), 3494-7.

Voorn, G. P., Thompson, J., Goessens, W. H., Schmal-Bauer, W. C., Broeders, P. H., and Michel, M. F. (1994). Paradoxical dose effect of continuously administered cloxacillin in treatment of tolerant Staphylococcus aureus endocarditis in rats. *J Antimicrob Chemother* 33(3), 585-93.

Woolfrey, B. F., and Enright, M. A. (1990). Ampicillin killing curve patterns for ampicillin-susceptible nontypeable Haemophilus influenzae strains by the agar dilution plate count method. *Antimicrob Agents Chemother* 34(6), 1079-87.

Woolfrey, B. F., Gresser-Burns, M. E., and Lally, R. T. (1987). Ampicillin killing curve patterns of Haemophilus influenzae type b isolates by agar dilution plate count method. *Antimicrob Agents Chemother* 31(11), 1711-7.

Yanagisawa, C., Hanaki, H., Matsui, H., Ikeda, S., Nakae, T., and Sunakawa, K. (2009). Rapid depletion of free vancomycin in medium in the presence of beta-lactam antibiotics and growth restoration in Staphylococcus aureus strains with beta-lactam-induced vancomycin resistance. *Antimicrob Agents Chemother* 53(1), 63-8.

Young, D., Hussell, T., and Dougan, G. (2002). Chronic bacterial infections: living with unwanted guests. *Nat Immunol* 3(11), 1026-32.

Zabransky, R. J., Johnston, J. A., and Hauser, K. J. (1973). Bacteriostatic and bactericidal activities of various antibiotics against Bacteroides fragilis. *Antimicrob Agents Chemother* 3(2), 152-6.

Zhao, X., and Drlica, K. (2001). Restricting the selection of antibiotic-resistant mutants: a general strategy derived from fluoroquinolone studies. *Clin Infect Dis* 33 Suppl 3, S147-56.

CHAPTER III

SMALL COLONY VARIANTS AND CHRONIC INFECTIONS

Section 1
Small Colony Variants: Current Knowledge

S mall colony variants (SCVs) constitute a naturally occurring, slow-growing subpopulation of bacteria that form small colonies (less than one-tenth of the size of parent colonies) on solid media (Proctor *et al.* 2006). SCVs were first described around 100 years ago and since then have been reported in a wide range of bacterial genera and species. However, they have been most extensively studied for *Staphylococcus aureus* (Proctor *et al.* 2006). The major characteristics of SCVs include slow growth rate, lack of pigmentation, reduced hemo-lytic activity, reduced coagulase activity, increased antibiotic re-sistance especially to aminoglycosides, reduced carbohydrate utiliza-tion and low virulence with reduced production of virulence factors (McNamara and Proctor 2000; Proctor *et al.* 2006). Since the growth rate of SCVs is approximately nine times lower than the parent strains (Proctor *et al.* 1998; Proctor *et al.* 2006), they require a longer incubation time (48-72 h) to form pinpoint colonies on agar. They are implicated in a number of chronic infections, especially cystic fibrosis (CF) and osteomyelitis. SCVs are difficult to recognize using routine laboratory biochemical tests because of their atypical mor-phological and physiological features, thus presenting a challenge to clinical microbiologists (von Eiff 2008).

SCVs can be selected by suboptimal concentrations of ami-noglycosides both *in vitro* and *in vivo* (Rusthoven *et al.* 1979; Mush-er *et al.* 1979; Wilson and Sanders 1976; Soren and Nilsson 1984). SCVs are also generated following exposure of bacterial cultures to metal ions, hydrogen peroxide, human serum and different types of antibiotics (Youmans and Delves 1942; Swingle 1935; Davies *et al.* 2007; Mitsuyama *et al.* 1997; Pan *et al.* 2002). Similarly, old cultures are also associated with the formation of SCVs (Swingle 1935).

The slow growth rate and formation of small colonies is often due to the inability of the bacteria to synthesize certain substances required for their growth; thus, supplementation of these substances in the growth medium can result in a normal growth rate (Proctor *et al.* 2006; von Eiff 2008). Most of the SCVs reported are auxotrophic for hemin, menadione or thiamine (Proctor et al. 2006), even though other compounds such as thymidine (Besier *et al.* 2007), unsaturated fatty acids (Kaplan and Dye 1976) and CO_2 (Sherris 1952) may also stimulate the growth of SCVs (Proctor *et al.* 2006).

Different types of SCVs that can be isolated under a variety of conditions are given below:

i. SCVs that revert on subculturing (Massey *et al.* 2001)

ii. Stable SCVs selected by antibiotics that revert in the presence of hemin, thiamine, menadione or thymidine (Proctor *et al.* 2006; Chatterjee *et al.* 2007, Besier *et al.* 2008)

iii. Those requiring unsaturated fatty acids for reversion (Kaplan and Dye 1976)

iv. Those induced by intracellular milieu (Vesga *et al.* 1996)

iv. *S. aureus* SCVs induced by *P. aeroginosa* or its exoproduct, 4-hydroxy-2-heptylquinoline-N-oxide (HQNO) (Hoffman *et al.* 2006)

v. SCVs due to mutations in F_oF_1-ATPase (Jensen and Michelsen 1992)

vi. Fried-egg SCVs of *S. aureus* that revert spontaneously to wild type in MH broth at a frequency of 10^{-3} (Kahl *et al.* 2003)

vii. *E. coli* SCVs that carry two independent mutations that helps to take up hemin and revert in the presence of hemin (Roggenkamp *et al.* 1998)

vii. Hemin-dependent *E. coli* SCVs that revert to normal growth on Schaedler's agar under anaerobic conditions (Tappe *et al.* 2006)

ix. SCVs generated by prolonged intracellular residence of *S. enterica* serovar *typhimurium* which are auxotrophic to δ-aminolevulinic acid or aromatic amino acids (Cano *et al.* 2003)

x. Rough small colony variants of *P. aeroginosa* with almost the same doubling time as the wild types (Drenkard and Ausubel 2002; Kirisits *et al.* 2005)

xi. Quinolone-induced SCVs, not auxotrophic to hemin, thiamine or thymidine (Pan *et al.* 2002)

xii. SCVs of *Burkholderia cepacia* complex those are not auxotrophic to hemin, thymidine, menadione or amino acids (Haussler *et al.* 2003)

xiii. SCVs of *P. aerogenosa* not auxotrophic to hemin, thymidine, menadione or amino acids (Haussler *et al.* 1999)

xiv. *S. aureus* SCVs selected by triclosan that are not auxotrophic for hemin, menadione, thiamine or thymidine (Seaman *et al.* 2007; Bayston *et al.* 2007) and are not reverted by repeated subculture on enriched medium (Bayston *et al.* 2007).

The reversion of SCVs to the wild phenotype occurs at various rates depending on the type of mutation (McNamara and Proctor 2000). SCVs generated after penicillin treatment may revert to the normal phenotype after transfer to fresh media (McNamara and Proctor 2000). On the other hand, stable SCVs are often formed following treatment with aminoglycosides (Aoki *et al.* 1998; Proctor *et al.* 2006). However, Massey *et al.* (2001) reported that the SCVs emerging after exposure to gentamicin result from phenotypic switching and that they can switch repeatedly between the two phenotypes (small and large colony types) when exposed to cycles of antibiotic treatment followed by growth in antibiotic-free medium.

Biochemical aspects of SCV formation

The phenotypic characteristics of SCVs can be related to defects in electron transport pathways. Defects in electron transport may affect the capacity of bacteria to produce ATP resulting in slow growth (McNamara and Proctor 2000). Genetic mutations in *hemB*, *menD*, *thyA* and *ctaA* produce the SCV phenotype (von Eiff *et al.* 1997b; Bates *et al.* 2003; Besier *et al.* 2007; Proctor *et al.* 2006). A mutation in *hemB* and *ctaA* blocks the biosynthesis of hemin, which is used in the cytochrome synthesis, whereas a mutation in *menD* blocks the synthesis of menadione, used in menaquinone synthesis (menadione is isoprenylated to menaquinone, which accepts electrons from NADH/FADH2 in the electron transport chain) (Proctor *et al.* 2006). Since both menaquinone and cytochromes are components of the electron transport system, these mutations result in defective electron transport (Proctor *et al.* 2006). Similarly, a mutation in *thyA*, which encodes thymidylate synthase, results in impaired thymidine metabolism leading to SCV formation (Proctor *et al.* 2006).

Since defining the precise genetic mutations in SCVs from clinical isolates is difficult, the physiological characteristics of SCVs are mainly studied using *hemB* (von Eiff *et al.* 1997b), *menD* (Bates *et al.* 2003) or *thyA* (Besier *et al.* 2007) mutants of *S. aureus*. All of these mutants show typical characteristics of SCVs such as slow growth, small colonies, decreased hemolytic and coagulase activity, reduced pigment formation and resistance to aminoglycosides. Complementing these mutants with *hemB*, *menD* or *thyA* respectively could restore the normal phenotype (von Eiff *et al.* 1997b; Bates *et al.* 2003; Besier *et al.* 2007). When subjected to phenotype microarray analysis, the *hemB* and *menD* mutants were found to be defective in utilizing a variety of carbon sources that generate ATP via electron transport, whereas compounds that provided ATP in the absence of electron transport were found to be stimulating the growth of these mutants (Kohler *et al.* 2003; von Eiff 2008; McNamara and Proctor 2000). *hemB* mutants upregulate the genes of enzymes involved in the glycolytic and fermentative pathways and arginine deaminase pathway that help to produce ATP (Kohler *et al.* 2003; Seggewiss *et al.* 2006). Since ATP is generated by F_oF_1-ATPase, an integral membrane protein complex that uses the energy of a transmembrane proton gradient to synthesize ATP, defects in F_oF_1-ATPase also generate SCVs (Proctor *et al.* 2006; Proctor 2000; McNamara and Proctor 2000). However, those SCVs may not be auxotrophic to hemin or menadione even though they have defects in electron transport (Proctor 2000; Proctor *et al.* 2006). Similarly, under anaerobic conditions, SCVs are generated due to their inability to synthesize menaquinone (Proctor *et al.* 2006; McNamara and Proctor 2000). However, in *E. coli*, anaerobic growth may not generate SCVs since they have two quinones, ubiquinone and menaquinone, the former used in the presence of oxygen and the latter used in anaerobic conditions (McNamara and Proctor 2000). Additionally, defects in unsaturated fatty acid biosynthesis may generate SCVs (Kaplan and Dye 1976). The formation of menaquinone from menadione requires the synthesis of unsaturated fatty acids in the form of an isoprenoid tail that is added to menadione to form menaquinone. Hence a defect in the biosynthesis of unsaturated fatty acids disrupts the electron transport chain (Kaplan and Dye 1976; McNamara and Proctor 2000). Thus, the above factors indicate that defects in electron transport are mainly responsible for the generation of SCVs.

Defects in electron transport result in reduced carotenoid biosynthesis, which leads to reduced pigment formation by SCVs (Proctor *et al.* 2006). Similarly, defects in electron transport decrease the amount of ATP used for cell-wall biosynthesis, leading to a slower growth rate and reduced membrane potential (Proctor *et al.* 2006). Reduction in the membrane potential results in the decreased uptake of cationic compounds including the aminoglycoside antibiotic (Proctor *et al.* 2006). The uptake of aminoglycosides depends on the membrane potential. Initiation of aminoglycoside uptake by bacteria requires a threshold level of membrane potential; above this level, the drug uptake is directly dependent on the magnitude of the membrane potential (Eisenberg *et al.* 1984). Since SCVs show reduced membrane potential, uptake of aminoglycosides will be low, resulting in reduced killing by the antibiotic. SCVs may be resistant to killing by other antibiotics such as penicillin, probably due to their slow growth rate (Proctor 2000; Eng *et al.* 1991; Gilbert *et al.* 1990; Brown *et al.* 1990).

Attachment of *S. aureus* to host cells involves many adhesion proteins and host cell proteins. SCVs show increased expression of clumping factors and fibronectin-binding proteins that aid in efficient host cell attachment (Vaudaux *et al.* 2002). Similarly, internalization and intracellular uptake of SCVs is more efficient when compared to the wild type strains (Vaudaux *et al.* 2002). Moreover, SCVs produce only low levels of alpha-toxins (Balwit *et al.* 1994; von Eiff *et al.* 1997b; von Eiff *et al.* 2001; Bates *et al.* 2003; Vaudaux *et al.* 2002; Kahl *et al.* 2005), which can be linked to decreased electron transport since electron transport inhibitors produce SCVs that are unable to hemolyze rabbit RBC (Proctor 2000). Alpha-hemolysin produced by the wild type *S. aureus* is cytotoxic to cells. Thus, a decrease in the alpha-toxin production by SCVs helps in their intracellular persistence (Vann and Proctor 1988; Balwit *et al.* 1994; von Eiff *et al.* 2001). The low levels of cytotoxin production by SCVs downregulate the induction of cell lysis or apoptosis (Proctor *et al.* 2006). When supplemented with hemin or menadione, SCVs revert to rapid growth which leads to decreased intracellular persistence (Balwit *et al.* 1994). Since the intracellular milieu protects bacteria from antibodies, complements and many antibiotics (Sendi and Proctor 2009), increased intracellular persistence of SCVs could be a survival strategy of *S. aureus* (Vesga *et al.* 1996; Proctor *et al.* 2006; Sendi and Proctor 2009).

Ultrastructure of SCVs

Under a scanning electron microscope, a normal *S. aureus* population is seen as homogenous cocci with no visible debris, whereas SCVs are seen as a heterogeneous population of different sizes, some of which are empty and covered with debris (Kahl *et al.* 2003a; Aoki *et al.* 1998; Wellinghausen *et al.* 2009). Normal bacteria are sphere shaped and thin-walled with the cytoplasm having areas of rough granular appearance interspersed with areas of fine granular appearance (Adler *et al.* 2005). SCVs, on the other hand, show thick walls and irregular cytoplasm with dense granular appearance at the periphery and fine granular materials at the center (Adler *et al.* 2005). Similarly, SCVs with incomplete, branched and multiple cross walls without regular cell separation are common (Proctor *et al.* 2006; Kahl *et al.* 2003a; Wellinghausen *et al.* 2009). In contrast, normal *S. aureus* show regular cell separation by cross walls (Bulger and Bulger 1967; Proctor *et al.* 2006; Kahl *et al.* 2003a; Wellinghausen *et al.* 2009). Some of the SCVs are nearly eight times larger than the normal cells which could be due to impaired cell separation (Kahl *et al.* 2003a; Seaman *et al.* 2007; Wellinghausen *et al.* 2009). Ghost (empty) cells with defective cell walls, indicative of dying cells, can also be seen (Adler *et al.* 2005; Aoki *et al.* 1998; Wellinghausen *et al.* 2009). All structural alterations can be reversed by supplementing the required auxotrophic agent (Kahl *et al.* 2003a; Proctor *et al.* 2006). However, some researchers could not find any significant differences between the SCVs and normal cells in the cell size or morphology by electron microscopy (Haussler *et al.* 1999; von Eiff *et al.* 1997b).

Role of SCVs in biofilms

SCVs may facilitate biofilm formation in chronic infections including CF, osteomyelitis, otitis media and device-related infections (Haussler *et al.* 2003; von Gotz *et al.* 2004; Al Laham *et al.* 2007; Fergie *et al.* 2004). The respiratory tract of CF patients may provide a unique environment for the selection of a subgroup of autoaggregative and highly adherent *P. aeruginosa* SCVs (Haussler *et al.* 2003; Singh *et al.* 2010). These SCVs are hyperpiliated and exhibit increased twitching motility and they have the capacity to form and persist within the biofilms, thus playing an important role in the pathogenesis of *P. aeruginosa* lung infection (Haussler *et al.* 2003; von Gotz *et al.* 2004; Ikeno *et al.* 2007). They also show upregulation of the type III protein secretion system, which can result in increased cytotoxicity for macrophages *in vitro* and increased virulence in mouse models of respiratory tract infection (von Gotz *et al.* 2004). SCVs isolated *in vitro* also display an autoaggregative phenotype, produce high amounts of polysaccharide intercellular adhesin and form highly structured biofilms (Singh *et al.* 2010; Al Laham *et al.* 2007).

Biofilms are responsible for some device-related infections (Proctor *et al.* 2006; von Eiff and Becker 2007). Frequent re-implantation failure of joint replacements infected with *P. aeruginosa* can be attributed to the increased slime or biofilm production on the antibiotic-loaded bone cement together with the formation of SCVs, which exhibit reduced susceptibility to antibiotics (Neut *et al.* 2005). Similarly, many pacemaker-related infections are associated with biofilms containing SCVs (Seifert *et al.* 2003; Seifert *et al.* 2005; von Eiff *et al.* 2005). In many cases, removal of the foreign material responsible for the infection is essential for the complete cure of device-related infections (Seifert *et al.* 2005; von Eiff *et al.* 1999).

Role of SCVs in persistent infections

SCVs can be selected by antibiotics, especially aminoglycosides, both *in vitro* and *in vivo* (Gerber and Craig 1982; Gerber *et al.* 1982). *S. aureus* SCVs selected *in vitro* by aminoglycosides are similar to the clinical isolates in that most of them are hemin or menadione auxotrophs (Proctor *et al.* 1998). Clinical cases have been reported wherein SCVs were isolated following aminoglycoside therapy (Seifert *et al.* 2003; Rolauffs *et al.* 2003; von Eiff *et al.* 1997a). Thus, within a host, antibiotic pressure may select electron transport deficient mutants which may survive intracellularly due to low levels of free hemin and menadione within the host cell (McNamara and Proctor 2000). When conditions become favorable, they may revert to a normal wild type population, thus resulting in recurrent infections (Proctor *et al.* 1995; Kahl *et al.* 1998). In most of the clinical cases, both the SCVs and large colony types have been isolated. Researchers have shown that both colony types are clonal indicating a common origin (Haussler *et al.* 1999; von Eiff *et al.* 1999; Sendi *et al.* 2006).

Isolation of SCVs from osteomyelitis and CF supports a pathogenic role for SCVs in such diseases (Proctor *et al.* 2006; Besier *et al.* 2007; Haussler *et al.* 1999; Haussler *et al.* 2003; Proctor *et al.* 1995; von Gotz *et al.* 2004). SCVs have also been isolated from device-related infections (von Eiff *et al.* 1999; Seifert *et al.* 2003), persistent wound infections (Abele-Horn *et al.* 2000) and persistent bovine mastitis (Atalla *et al.* 2008; Brouillette *et al.* 2004).

The rate of occurrence of SCVs may vary depending on the clinical conditions (Proctor *et al.* 2006). SCVs are detected in approximately 1% of isolates in a general microbiology laboratory, but their number is far greater in patients with CF and osteomyelitis (Proctor *et al.* 2006). In one study, 4 out of 14 patients with osteomyelitis, colonized with *S. aureus*, showed SCVs in their samples (von Eiff *et al.* 1997a). Similarly, SCVs were isolated from 24 out of 72 patients with CF, colonized with *S. aureus* (Kahl *et al.* 2003b). In other studies, SCVs were isolated from 38% of *P. aerogenosa*-positive CF patients

(Haussler *et al.* 1999), whereas 8 out of 19 CF patients with *Burkholderia cepacia*-like organisms harbored SCVs in their clinical samples (Haussler *et al.* 2003).

In patients with osteomyelitis, surgical placement of gentamicin beads along with debridement is a common practice of treatment. However, it is possible that the slow release of gentamicin by the beads may select SCVs. When the bone specimens from osteomyelitis patients were screened for *S. aureus* SCVs, SCVs were recovered only from those patients who had been on treatment with gentamicin beads, whereas no SCVs were recovered from patients who had not received gentamicin beads (von Eiff *et al.* 1997a). Moreover, those patients harboring SCVs had recurrent disease in spite of antibiotic therapy, whereas those with normal *S. aureus* only completely recovered following the antibiotic treatment. It is suggested that clinicians should check for the presence of SCVs in recurrent osteomyelitis after gentamicin bead treatment (Proctor *et al.* 1998).

SCVs in fatal infections

SCVs may be responsible for some fatal infections as well. A fatal infection due to SCVs of methicillin-resistant *S. aureus* in a patient with AIDS has been reported (Seifert *et al.* 1999). Similarly, SCVs of the *B. cepacia* complex has been reported to be responsible for the fatal outcome of lung transplantation in CF patients (Haussler *et al.* 2003). In another case, *S. aureus* SCVs were isolated from the blood cultures of a patient with acute leukemia and who was on treatment with vancomycin for a catheter-associated bloodstream infection (Adler *et al.* 2003). Even though the catheter was removed and adequate antibiotic therapy was given, the patient did not survive. Similarly, *S. epidermidis* SCVs were associated with the pathogenesis of intracardiac infections since they were isolated from a patient who ultimately died from prosthetic heart valve endocarditis in spite of antibiotic treatment (Baddour and Christensen 1987).

Virulence of SCVs in nematode and animal models

Reports on the virulence of SCVs have given contrasting results. Some experiments using animal models have shown that SCVs are less virulent than normal strains (Wise and Spink 1954; Swingle 1935; Musher *et al.* 1977; Pelletier *et al.* 1979). However, other studies have indicated that they are as virulent as parent strains (Musher *et al.* 1979; Bates *et al.* 2003). In a murine model of septic arthritis, *S. aureus* *hemB* mutants were found to produce severe arthritis with a high frequency than when inoculated with the wild type parent strain (Jonsson *et al.* 2003). They also produced almost 20 times more proteases *in vitro* than the parental strain. However, mice inoculated with mutants had a significantly lower bacterial burden in their kidneys and joints. In another study, *S. aureus hemB* mutants were found to be as virulent as the wild type whereas the *menD* mutants exhibited reduced bacterial densities in vegetations, kidneys and spleen (Bates *et al.* 2003). Recently, using *Caenorhabditis elegans* as an infection model, Sifri *et al.* (2006) reported that the clinical SCVs as well as *hemB* and *menD* mutants of *S. aureus* exhibited greatly reduced virulence when compared to the wild type parent strains; this reduction in virulence was not due to its impaired ability to colonize the nematode digestive tract.

Treatment of infections associated with SCVs

Only limited information is available regarding the treatment of infections associated with SCVs. Since aminoglycosides can select SCVs, they may not be suitable for the treatment of infections caused by SCVs. Similarly, β-lactam antibiotics may not be effective probably due to the slow growth of SCVs. Vancomycin had approximately 50% less activity against *hemB* mutants which was attributed to the unique expression patterns in SCVs (Tsuji *et al.* 2008). However, daptomycin achieved bactericidal activity in a concentration-dependent manner against both *S. aureus hemB* mutants and their parental strains (Begic *et al.* 2009). Similarly, the maximum killing effect was found to be reduced by 7.7-fold for vancomycin whereas it was reduced only by 1.5-fold for daptomycin against *S. epidermidis* SCV (Wu *et al.* 2009). Thus, daptomycin may have the potential to be used as a therapeutic agent against infections associated with SCVs. Some drug combinations may also be effective against such infections (Proctor *et al.* 1995).

Conclusion

SCVs constitute a slow-growing subpopulation of bacteria that form colonies nearly one-tenth of the size of normal wild-type bacteria. They are electron transport deficient mutants and are generally auxotrophic to hemin, thiamine or menadione. They have reduced membrane potential and can be easily selected by aminoglycosides both *in vitro* and *in vivo*. SCVs are considered responsible for many chronic and fatal infections. Because of their atypical growth characteristics, they are not easily identified in routine laboratory tests and hence present a challenge to clinical microbologists.

Section 2
Small Colony Variants:
Are They Responsible for Chronic Infections?

A large volume of scientific literature implicates SCVs as important etiological agents responsible for many chronic infections. However, there are many reasons to challenge this assumption, which are given below.

1. The current definition for SCV is not appropriate.
SCVs are defined as bacterial variants that form very small colonies (less than one-tenth of the size of normal colonies). By this definition, bacteria in a population are divided into two groups based on the size of colonies; i.e., normal sized colonies and SCVs. However, colonies of many different sizes can be isolated by incubating a bacterial culture with different concentrations of aminoglycosides. The current definition excludes small colonies greater than one-tenth of the size of parent colony. Similarly it categorizes all colonies smaller than one-tenth of the size of parent colony into a single group.

Brouillette *et al.* (2004) isolated SCVs of *S. aureus* Newbould *hemB* mutants after 48 h of incubation at 37°C on TSA. Those colonies were approximately 1 mm in diameter, whereas the parent colonies were 4 mm or larger in diameter. In this case, *hemB* mutants which were only around one-quarter the size of parent colonies were considered as SCVs. Similarly, SCVs of *Proteus mirabilis* isolated after treatment with genatmicin at a concentration of 7.5 µg/ml were 0.35 mm in diameter whereas the parent strain was 1.19 mm (Musher *et al.* 1979). On the other hand, Kaplan and Dye (1976) reported that the size of *S. aureus* SCVs ranged from 0.1 to 0.01 mm in diameter whereas the parent colonies ranged from 2-3 mm diameter. Thus, among SCVs, a ten-fold difference in the size of colonies was noticed. The current definition of SCVs does not explain why there are colonies of different sizes and how SCVs of ten-fold variation in size differ from one another.

2. There is no advantage in generating high frequencies of hemin-deficient mutants among the *Enterobacteriacae* family.
It is argued that SCV generation is a survival strategy of bacteria to resist adverse conditions, especially antibiotic therapy. However, if the formation of SCV is a survival strategy, why do high frequencies of hemin-deficient mutants occur among the *Enterobacteriacae* family? Similarly what is the fate of those hemin mutants of *Enterobacteriacae*? It is documented that *Enterobacteriacae* lacks the ability to take up hemin (Sasarman *et al.* 1968). To revert to the normal wild type, it needs a second independent mutation that helps it take up hemin (Roggenkamp *et al.* 1998). However, the frequency of this second mutation is very low (Roggenkamp *et al.* 1998). This would mean that the hemin-deficient mutants of *Enterobacteriacae* will remain as SCVs even in the presence of hemin, thus offering them no growth advantages.

3. Do all SCVs that revert in the presence of hemin/menadione carry mutations in the hemin/menadione biosynthetic pathway? What are the conditions under which reversion occurs?
SCVs are considered to be electron deficient mutants. However, in many reported cases, SCVs were only assumed to be mutants based on their ability to revert in the presence of hemin, menadione or thymidine (Proctor *et al.* 1995; Haussler *et al.* 2003; Tappe *et al.* 2006; von Eiff *et al.* 1999; Abele-Horn *et al.* 2000; Seifert *et al.* 1999). It is not known whether they carry any mutations in the hemin/menadione/thymidine biosynthetic pathways.

McNamara and Proctor (2000) postulated that antibiotics select electron-deficient mutants that may persist intracellularly due to low levels of free hemin or menadione within the host cells. If free hemin and menadione levels inside the cells are low, what are the conditions under which the reversion occurs *in vivo*?

4. SCVs may not be responsible for chronic infections.
There are a number of case reports indicating the association of SCVs in chronic infections (reviewed earlier). However, it has not been conclusively proved that these infections are indeed caused by SCVs. In most of the clinical cases reported, the primary culture contained a mixture of normal colonies and SCVs. Even though they were demonstrated to be clonal, none of the reports have shown that the normal colonies were the revertants of SCVs.

It can be proposed that SCVs are responsible for chronic infections only if:

1) SCVs, by themselves, are directly responsible for infections. However, SCVs are less virulent and, in most cases, they are co-cultured with normal forms (Wise and Spink 1954; Swingle 1935; Musher *et al.* 1977; Pelletier *et al.* 1979; Sifri *et al.* 2006).
2) SCVs can revert to normal virulent colonies *in vivo* (however, there are no indications that the large colony types co-cultured with SCVs are the revertants of SCVs).

Normal colony type

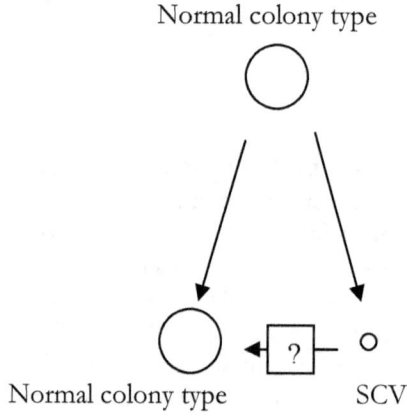

Normal colony type   SCV

Figure 1. Clonality of normal colony types and SCVs isolated from clinical cases. Even though both forms are clonal, there is no indication that the large colony types are the revertants of SCVs.

As shown in Fig. 1., a normal colony on multiplication can give rise to both normal and small colony forms. Researchers assume that large colony forms are the revertants of SCVs. However, reversion of SCVs to normal colony forms *in vivo* has not been proved. Moreover, SCVs can remain viable in animal tissues without any signs of infection (Wise and Spink 1954). If the bacteria that form normal colonies are not the revertants of SCVs, how can it be postulated that the SCVs are responsible for chronic infections? Rather, the infection can be due to normal bacteria itself that somehow persist under the disease conditions, whereas SCVs may remain without causing any infections due to their low virulence.

Recently, Chatterjee *et al.* (2008) isolated thymidine-dependent (TD) small colony variants (TD-SCVs) from CF patients and sequenced *thyA* in clinical TD-SCVs and compared it with isogenic normal *S. aureus* strains. This was the only article found that gave a detailed genetic analysis of both SCVs and their isogenic normal strains. In two of the normal strains, the *thyA* sequence was identical to that of reference strain N315 and had no mutations. Other normal strains carried one to three non-synonymous point mutations. (However, they exhibited normal growth in spite of carrying a few point mutations). All the SCVs carried different mutations, including non-synonymous point mutations, two 3-bp in-frame deletions, six 1-bp deletions, an 11-bp deletion resulting in a frame shift mutation and a point mutation resulting in a stop mutation. A chart showing mutations in *thyA* in TD-SCVs and their isogenic normal strains (Chatterjee *et al.* 2008) is given below. Even though non-synonymous point mutations were detected in SCVs, all of them were different from the point mutations noticed in the isogenic normal strains. Similarly, mutations other than non-synonymous point mutations found in SCVs were not detected in normal strains. This indicates that those normal strains were not the revertants of the isogenic SCVs even though both were clonal. Since normal colonies are more virulent than SCVs, the chronic infection can be due to those normal strains itself and not due to SCVs.

Mutations in *thyA* in TD-SCVs and its isogenic normal strains (Chatterjee et al. 2008)	
Parent strain	Thymidine-dependent SCV
Nil	3-bp deletion
Nil	3-bp deletion
Point mutation Cys →Arg	Point mutations Leu →Por, Gly →Ser, Asn →Asp, 1-bp deletion →stop codon
Point mutations Ser →Ala and Gly →Ser	11-bp deletion →frameshift mutation → truncated protein
Point mutations Gly →Asp and Leu → Phe	1-bp deletion →frameshift mutation → stop codon
Point-mutations Ser →Phe, Asp →Gly and Val →Ile	Point-mutation Arg →Gln, and 1-bp deletion →frameshift mutation →stop codon
Point-mutations Tyr →Asn, Tyr →Ile, Gly →Asp and 1-bp deletion → frameshift mutation →stop codon	Point-mutation Tyr →Asn and point mutation →stop codon
1-bp deletion →frameshift mutation → stop codon	7-bp deletion →frameshift mutation → stop codon

Table 1. Mutations in *thyA* in TD-SCVs and its isogenic normal strains isolated from the airway secretions of CF patients (Chatterjee *et al.* 2008)

SCVs can be isolated after gentamicin treatment both *in vitro* and *in vivo*, and most of them are reported to be hemin or menadione auxotrophs. Von Eiff *et al.* (1997) isolated SCVs of *S. aureus* from osteomyelitis patients who had received gentamicin beads. They selected 14 patients with osteomyelitis, of which 4 had received previous gentamicin bead treatment and 10 had not. SCVs were isolated from all 4 patients who had received beads; 3 of these patients showed both small and large colony types, whereas the fourth patient had only SCVs. Those 10 patients without bead treatment did not have a relapse of the disease once the optimal treatment was given. However, for the 4 patients, treatment with antibiotics failed despite the parent *S. aureus* strain being sensitive to antibiotics *in vitro*. This clinical trial may be indicative of SCVs being responsible for chronic osteomyelitis. However, authors themselves admitted that the treatment failure could not be attributed to SCVs alone because the entry criteria included only those patients with positive *S. aureus* cultures and whose gentamicin bead therapy had failed. Similarly it should be noticed that in 3 patients, both large and small colonies were recovered. Unless it is proved that the large colonies are the revertants of SCVs, it cannot be concluded that SCVs are responsible for infections. Moreover, debridement along with gentamicin bead treatment is a widely accepted method with good clinical success rates in the treatment of osteomyelitis (Mendel *et al.* 2005; Shih *et al.* 2005; Joosten *et al.* 2004; Klemm 2001). Hence, SCVs may not be responsible for chronic infections since gentamicin bead treatment has good success rates despite the selection of SCVs at high frequencies due to the slow release of gentamicin.

5. SCVs may not be responsible for fatal infections.
To my knowledge there are four reports wherein SCVs are considered responsible for fatal infections. Seifert *et al.* (1999) reported a fatal case due to SCVs of methicillin-resistant *S. aureus* (MRSA) in an AIDS patient. This was the first reported case of fatal infection due to SCVs. In this report, a patient diagnosed with AIDS in 1986 and who had a medical history of *Pneumocystis carinii* pneumonia, tuberculosis, recurrent oral thrush and recurrent abscess of the right hip had a traffic accident in 1996 that resulted in severe cerebral trauma. Blood cultures taken were positive for *S. aureus* SCVs. The *S. aureus* normal phenotype was isolated from the nose and throat specimen but not from the blood cultures. Even though systemic antibiotic therapy was given, the patient died of refractory septic shock after

the sixth day of admission. Post-mortem revealed a deep-seated abscess of the right hip and osteomyelitis, and the cultures taken from the bone and abscess showed both large and small colonies of *S. aureus*. SCVs were auxotrophic for thymidine or menadione and were clonal with large colonies (here again, the authors have not shown whether the normal large colonies were the revertants of SCVs). Because SCVs were isolated from blood cultures, bone and abscess specimens, the authors claimed that SCVs could be responsible for the patient's death. However, in this case, the picture is much more complicated because the patient with a long medical history of many infections had severe cerebral trauma following the traffic accident. He died six days after the accident; the immediate cause of death, therefore, can be attributed to the accident only rather than to SCVs or even *S. aureus* itself, since the deep-seated abscess and osteomyelitis harboring large and small colony forming *S. aureus* might have developed long before his admission to the hospital. Considering the fact that he had severe cerebral trauma with hemiparesis and seizures and a long history of chronic infections, it was too early to blame SCVs for his immediate death.

Another case of fatal outcome due to SCVs was reported in CF patients after lung transplantation (Haussler *et al.* 2003). Out of the 470 CF patients screened for *Burkholderia cepacia*-like organisms, 19 patients were positive for the organism and 8 harbored SCVs. Due to severe lung disease, 3 of them received bilateral lung transplantation. Before the transplantation, the respiratory tract specimen from one of them was positive for both the *B. multivorans* wild type and SCVs; the second transplantation candidate was positive for both *B. cepacia* genomovar III wild type and SCVs; and the third candidate for the *B. multivorans* wildtype only, with no SCVs. The first patient died 4 days following transplantation due to sepsis. Only SCVs and no wild type *B. multivorans* were recovered from autopsy materials. The second patient had complications after the first transplantation; this patient underwent re-transplantation after 4 months and died several days later. Here also, only SCVs and no wild type *B. cepacia* genomovar III were recovered from empyema and blood cultures. The third patient who harbored only the wild types of *B. multivorans* had no complications after transplantation. Since SCVs and no wild types were recovered after transplantation from the above two patients, the authors concluded that SCVs were responsible for their deaths since they escaped antibiotic killing and exhibited increased serum resistance. However, just because SCVs were recovered from different autopsy specimens and were resistant to the bactericidal activity of serum does not indicate that they were responsible for the fatalities. It is also not known whether the fatal outcome was host-related or pathogen-related. Even if it was pathogen-related, how did the authors conclude that it was due to B. *cepacia*-like organisms? The authors had only screened the CF patients for B. *cepacia*-like organisms, and there is no evidence that the cause of death was indeed due to the B. *cepacia* complex itself, even if the cause was pathogen-related.

In the third case, the blood cultures of a patient who was undergoing treatment for acute myeloid leukemia yielded S. *epidermidis* that were methicillin-resistant but were susceptible to vancomycin and teicoplanin (Adler *et al.* 2003). After starting treatment with vancomycin, both normal forms and SCVs of S. *epidermidis* were isolated from the blood culture. Both colony types were susceptible to vancomycin, but SCVs were resistant to teicoplanin, whereas the normal colony forms were not. In spite of treatment with amikacin and rifampicin, the patient died on day 20 after the start of antibiotic therapy. There were no signs of septic thrombophlebitis or endocarditis. On histologic examination, the lower right lobe of the lung revealed multiple nodes with fungal elements. The authors concluded that persistence of the bloodstream infection was caused by the SCVs. Here again, there are no indications of SCVs being responsible for the infection. The authors ignored the role of normal colony forms, assuming that they were reverted from SCVs. Additionally, the role of fungal elements in the lung is also not known.

In the fourth reported case, a patient with aortic stenosis and a calcified aortic valve had undergone cardiac catheterization 5 weeks before hospitalization (Baddour and Christensen 1987). The left heart valve was hypokinetic and dilated and the patient had a history of heart murmur. Three days after admission, the patient underwent replacement of both aortic and mitral valves and was under intravenous and oral antibiotic therapy including cefamandole, vancomycin, rifampin and gentamicin. Due to separation of the posterior mitral leaflet from the annular ring, another surgery for mitral valve replacement was done, but the patient died. Blood and vegetation culture revealed both large and small colony forms of S. *epidermidis*. The authors suggested that SCVs were responsible for the intracardiac infections.

In all four cases, it can be noted that the patients had received antibiotic therapy for a long time. Moreover, they had long and complicated clinical histories. Antibiotic therapy might have selected SCVs, and both large and small colony forms were cultured in all cases. There are no indications that SCVs are responsible for the fatal infections in any of the above cases.

6. A switching mechanism between the normal bacterial population and SCVs may not exist.

Massey *et al.* (2001) proposed that SCVs could emerge by switching from the wild type. They detected SCVs of *S. aureus* after just 30 min of gentamicin exposure. Their number increased as the exposure time increased and reached a maximum by 14 h, but subsequently declined due to the emergence and overgrowth of gentamicin-resistant wild type bacteria. The authors hypothesized that the increase in the frequency of SCVs after gentamicin treatment for the first 14 h was either due to the very short generation time of SCVs or due to the switching from wild type to SCVs. The former hypothesis was ruled out since the actual mean generation time was found to be much higher than the wild type. To determine whether the emergence of SCVs depended on the initial inoculum size, they reduced the initial inoculum size of bacteria by 100-fold and calculated the percentage of SCVs. They found that the proportion of SCVs at 24 h increased when the initial inoculum size was reduced. They calculated that, if the emergence of SCVs actually depended on the initial inoculum, their number should have reduced at 24 h at low inoculum. Since they found an increased percentage of SCVs, it was concluded that the emergence of SCVs was not dependent on its initial numbers, but was due to the switching from the wild type bacteria.

However, the increase in the number of SCVs at a low inoculum size may not be due to the switching of wild type bacteria. Aminoglycosides selectively kill fast-dividing bacteria that have high membrane potential, leaving the slow-dividing SCVs. Thus they kill normal bacteria, whereas SCVs are spared and may gradually multiply and increase in number. Thus, until 14 h (in the above experiment), the increase in the population of SCVs could be due to the selective killing of normal bacteria along with the gradual multiplication of SCVs. But, at a high inoculum size, antibiotics may not kill all normal bacteria. Some bacteria may escape killing and may remain dormant for a short period of time but later may undergo adaptation and regrow and overcome the SCV population. On the other hand, at a low inoculum size, most of the normal bacteria get killed and thus the percentage of SCVs may increase. The switching mechanism proposed by Massey *et al.* (2001) could have resulted from the selective and complete killing of normal bacteria at a lower inoculum size, resulting in the selective multiplication and the increase in the number of SCVs.

Conclusions

SCVs can be isolated *in vitro* and *in vivo* under a variety of conditions, including antibiotic pressure. Various reports indicate that they are electron-deficient mutants. However, many SCVs do not revert in the presence of hemin, menadione, thiamine or thymidine and thus may not be auxotrophic mutants. It is also not known whether SCVs can revert to normal colony forms *in vivo*. Even though they can be isolated from many chronic infections, their role in causing these infections has not been proven conclusively. In most of the reported cases, SCVs were co-cultured with normal colony forms. Since there are no indications that the large colony forms are the revertants of SCVs, chronic infections could be due to the normal colony forms only that have persisted under those conditions. SCVs may not be responsible for fatal infections; reports associating SCVs with fatal infections are highly questionable. SCVs can be part of the normal life cycle of bacteria and may not have much clinical significance. However, the isolation of SCVs *in vivo* following antibiotic therapy may be indicative of the failure of antibiotics to reach throughout the site of infection at optimal concentrations, which may result in the survival of some normal bacteria and the selection of SCVs. The surviving normal bacteria may re-grow once the antibiotic pressure is removed, whereas the SCVs may remain without causing any infection. This aspect will be discussed in later sections.

References

Abele-Horn, M., Schupfner, B., Emmerling, P., Waldner, H., and Goring, H. (2000). Persistent wound infection after herniotomy associated with small-colony variants of Staphylococcus aureus. *Infection* 28(1), 53-4.

Adler, H., Schraner, E. M., Frei, R. and Wild, P. (2005). Ultrastructureof a clinical isolate of *Staphylococcus aureus* small colony variant and its revertant. *Microsc Microanal* 11 (Suppl 2), 982-983.

Adler, H., Widmer, A., and Frei, R. (2003). Emergence of a teicoplanin-resistant small colony variant of Staphylococcus epidermidis during vancomycin therapy. *Eur J Clin Microbiol Infect Dis* 22(12), 746-8.

Al Laham, N., Rohde, H., Sander, G., Fischer, A., Hussain, M., Heilmann, C., Mack, D., Proctor, R., Peters, G., Becker, K., and von Eiff, C. (2007). Augmented expression of polysaccharide intercellular adhesin in a defined Staphylococcus epidermidis mutant with the small-colony-variant phenotype. *J Bacteriol* 189(12), 4494-501.

Aoki, Y., Yamauchi, Y., Hayashi, H., Takayama, Y. and Tsuji, A. (1998). Characterization of small colony variants of methicillin-resistant *Staphylococcus aureus* regrown in the presence of arbekacin *Journal of Infection and Chemotherapy* 4(3), 107-111.

Atalla, H., Gyles, C., Jacob, C. L., Moisan, H., Malouin, F., and Mallard, B. (2008). Characterization of a Staphylococcus aureus small colony variant (SCV) associated with persistent bovine mastitis. *Foodborne Pathog Dis* 5(6), 785-99.

Baddour, L. M., and Christensen, G. D. (1987). Prosthetic valve endocarditis due to small-colony staphylococcal variants. *Rev Infect Dis* 9(6), 1168-74.

Balwit, J. M., van Langevelde, P., Vann, J. M., and Proctor, R. A. (1994). Gentamicin-resistant menadione and hemin auxotrophic Staphylococcus aureus persist within cultured endothelial cells. *J Infect Dis* 170(4), 1033-7.

Bates, D. M., von Eiff, C., McNamara, P. J., Peters, G., Yeaman, M. R., Bayer, A. S., and Proctor, R. A. (2003). Staphylococcus aureus menD and hemB mutants are as infective as the parent strains, but the menadione biosynthetic mutant persists within the kidney. *J Infect Dis* 187(10), 1654-61.

Bayston, R., Ashraf, W., and Smith, T. (2007). Triclosan resistance in methicillin-resistant Staphylococcus aureus expressed as small colony variants: a novel mode of evasion of susceptibility to antiseptics. *J Antimicrob Chemother* 59(5), 848-53.

Begic, D., von Eiff, C., and Tsuji, B. T. (2009). Daptomycin pharmacodynamics against Staphylococcus aureus hemB mutants displaying the small colony variant phenotype. *J Antimicrob Chemother* 63(5), 977-81.

Besier, S., Ludwig, A., Ohlsen, K., Brade, V., and Wichelhaus, T. A. (2007). Molecular analysis of the thymidine-auxotrophic small colony variant phenotype of Staphylococcus aureus. *Int J Med Microbiol* 297(4), 217-25.

Brouillette, E., Martinez, A., Boyll, B. J., Allen, N. E., and Malouin, F. (2004). Persistence of a Staphylococcus aureus small-colony variant under antibiotic pressure *in vivo*. *FEMS Immunol Med Microbiol* 41(1), 35-41.

Brown, M. R., Collier, P. J., and Gilbert, P. (1990). Influence of growth rate on susceptibility to antimicrobial agents: modification of the cell envelope and batch and continuous culture studies. *Antimicrob Agents Chemother* 34(9), 1623-8.

Bulger, R. J., and Bulger, R. E. (1967). Ultrastructure of small colony variants of a methicillin-resistant Staphylococcus aureus. *J Bacteriol* 94(4), 1244-6.

Cano, D. A., Pucciarelli, M. G., Martinez-Moya, M., Casadesus, J., and Garcia-del Portillo, F. (2003). Selection of small-colony variants of Salmonella enterica serovar typhimurium in nonphagocytic eucaryotic cells. *Infect Immun* 71(7), 3690-8.

Chatterjee I, Herrmann, M., Proctor, R. A., Peters, G. and Kahl, B. C. (2007) Enhanced post-stationary-phase survival of a clinical thymidine-dependent small-colony variant of Staphylococcus aureus results from lack of a functional tricarboxylic acid cycle. *J Bacteriol* 189, 2936-2940.

Chatterjee I, Kriegeskorte, A., Fischer, A., Deiwick, S., Theimann, N., Proctor, R. A., Peters, G., Herrmann, M. and Kahl, B. C. (2008) In vivo mutations of thymidylate synthase (encoded by thyA) are responsible for thymidine dependency in clinical small-colony variants of Staphylococcus aureus. *J Bacteriol* 190, 834-842.

Davies, J. A., Harrison, J. J., Marques, L. L., Foglia, G. R., Stremick, C. A., Storey, D. G., Turner, R. J., Olson, M. E., and Ceri, H. (2007). The GacS sensor kinase controls phenotypic reversion of small colony variants isolated from biofilms of Pseudomonas aeruginosa PA14. *FEMS Microbiol Ecol* 59(1), 32-46.

Drenkard, E., and Ausubel, F. M. (2002). Pseudomonas biofilm formation and antibiotic resistance are linked to phenotypic variation. *Nature* 416(6882), 740-3.

Eisenberg, E. S., Mandel, L. J., Kaback, H. R., and Miller, M. H. (1984). Quantitative association between electrical potential across the cytoplasmic membrane and early gentamicin uptake and killing in Staphylococcus aureus. *J Bacteriol* 157(3), 863-7.

Eng, R. H., Padberg, F. T., Smith, S. M., Tan, E. N., and Cherubin, C. E. (1991). Bactericidal effects of antibiotics on slowly growing and nongrowing bacteria. *Antimicrob Agents Chemother* 35(9), 1824-8.

Fergie, N., Bayston, R., Pearson, J. P., and Birchall, J. P. (2004). Is otitis media with effusion a biofilm infection? *Clin Otolaryngol Allied Sci* 29(1), 38-46.

Gerber, A. U., and Craig, W. A. (1982). Aminoglycoside-selected subpopulations of Pseudomonas aeruginosa: characterization and virulence in normal and leukopenic mice. *J Lab Clin Med* 100(5), 671-81.

Gerber, A. U., Vastola, A. P., Brandel, J., and Craig, W. A. (1982). Selection of aminoglycoside-resistant variants of Pseudomonas aeruginosa in an *in vivo* model. *J Infect Dis* 146(5), 691-7.

Gilbert, P., Collier, P. J., and Brown, M. R. (1990). Influence of growth rate on susceptibility to antimicrobial agents: biofilms, cell cycle, dormancy, and stringent response. *Antimicrob Agents Chemother* 34(10), 1865-8.

Haussler, S., Lehmann, C., Breselge, C., Rohde, M., Classen, M., Tummler, B., Vandamme, P., and Steinmetz, I. (2003). Fatal outcome of lung transplantation in cystic fibrosis patients due to small-colony variants of the Burkholderia cepacia complex. *Eur J Clin Microbiol Infect Dis* 22(4), 249-53.

Haussler, S., Tummler, B., Weissbrodt, H., Rohde, M., and Steinmetz, I. (1999). Small-colony variants of Pseudomonas aeruginosa in cystic fibrosis. *Clin Infect Dis* 29(3), 621-5.

Hoffman, L. R., Deziel, E., D'Argenio, D. A., Lepine, F., Emerson, J., McNamara, S., Gibson, R. L., Ramsey, B. W., and Miller, S. I. (2006). Selection for Staphylococcus aureus small-colony variants due to growth in the presence of Pseudomonas aeruginosa. *Proc Natl Acad Sci U S A* 103(52), 19890-5.

Ikeno, T., Fukuda, K., Ogawa, M., Honda, M., Tanabe, T., and Taniguchi, H. (2007). Small and rough colony pseudomonas aeruginosa with elevated biofilm formation ability isolated in hospitalized patients. *Microbiol Immunol* 51(10), 929-38.

Jensen, P. R., and Michelsen, O. (1992). Carbon and energy metabolism of atp mutants of Escherichia coli. *J Bacteriol* 174(23), 7635-41.

Jonsson, I. M., von Eiff, C., Proctor, R. A., Peters, G., Ryden, C., and Tarkowski, A. (2003). Virulence of a hemB mutant displaying the phenotype of a Staphylococcus aureus small colony variant in a murine model of septic arthritis. *Microb Pathog* 34(2), 73-9.

Joosten, U., Joist, A., Frebel, T., Brandt, B., Diederichs, S., and von Eiff, C. (2004). Evaluation of an in situ setting injectable calcium phosphate as a new carrier material for gentamicin in the treatment of chronic osteomyelitis: studies *in vitro* and *in vivo*. *Biomaterials* 25(18), 4287-95.

Kahl, B., Herrmann, M., Everding, A. S., Koch, H. G., Becker, K., Harms, E., Proctor, R. A., and Peters, G. (1998). Persistent infection with small colony variant strains of Staphylococcus aureus in patients with cystic fibrosis. *J Infect Dis* 177(4), 1023-9.

Kahl, B. C., Belling, G., Becker, P., Chatterjee, I., Wardecki, K., Hilgert, K., Cheung, A. L., Peters, G., and Herrmann, M. (2005). Thymidine-dependent Staphylococcus aureus small-colony variants are associated with extensive alterations in regulator and virulence gene expression profiles. *Infect Immun* 73(7), 4119-26.

Kahl, B. C., Belling, G., Reichelt, R., Herrmann, M., Proctor, R. A., and Peters, G. (2003a). Thymidine-dependent small-colony variants of Staphylococcus aureus exhibit gross morphological and ultrastructural changes consistent with impaired cell separation. *J Clin Microbiol* 41(1), 410-3.

Kahl, B. C., Duebbers, A., Lubritz, G., Haeberle, J., Koch, H. G., Ritzerfeld, B., Reilly, M., Harms, E., Proctor, R. A., Herrmann, M., and Peters, G. (2003b). Population dynamics of persistent Staphylococcus aureus isolated from the airways of cystic fibrosis patients during a 6-year prospective study. *J Clin Microbiol* 41(9), 4424-7.

Kaplan, M. L., and Dye, W. E. (1976). Growth Requirements of Some Small-Colony-Forming Variants of Staphylococcus-Aureus. *Journal of Clinical Microbiology* 4(4), 343-348.

Kirisits, M. J., Prost, L., Starkey, M., and Parsek, M. R. (2005). Characterization of colony morphology variants isolated from Pseudomonas aeruginosa biofilms. *Appl Environ Microbiol* 71(8), 4809-21.

Klemm, K. (2001). The use of antibiotic-containing bead chains in the treatment of chronic bone infections. *Clin Microbiol Infect* 7(1), 28-31.

Kohler, C., von Eiff, C., Peters, G., Proctor, R. A., Hecker, M., and Engelmann, S. (2003). Physiological characterization of a heme-deficient mutant of Staphylococcus aureus by a proteomic approach. *J Bacteriol* 185(23), 6928-37.

Massey, R. C., Buckling, A., and Peacock, S. J. (2001). Phenotypic switching of antibiotic resistance circumvents permanent costs in Staphylococcus aureus. *Curr Biol* 11(22), 1810-4.

McNamara, P. J., and Proctor, R. A. (2000). Staphylococcus aureus small colony variants, electron transport and persistent infections. *Int J Antimicrob Agents* 14(2), 117-22.

Mendel, V., Simanowski, H. J., Scholz, H. C., and Heymann, H. (2005). Therapy with gentamicin-PMMA beads, gentamicin-collagen sponge, and cefazolin for experimental osteomyelitis due to Staphylococcus aureus in rats. *Arch Orthop Trauma Surg* 125(6), 363-8.

Mitsuyama, J., Yamada, H., Maehana, J., Fukuda, Y., Kurose, S., Minami, S., Todo, Y., Watanabe, Y., and Narita, H. (1997). Characteristics of quinolone-induced small colony variants in Staphylococcus aureus. *J Antimicrob Chemother* 39(6), 697-705.

Musher, D. M., Baughn, R. E., and Merrell, G. L. (1979). Selection of small-colony variants of Enterobacteriaceae by *in vitro* exposure to aminoglycosides: pathogenicity for experimental animals. *J Infect Dis* 140(2), 209-14.

Musher, D. M., Baughn, R. E., Templeton, G. B., and Minuth, J. N. (1977). Emergence of variant forms of Staphylococcus aureus after exposure to gentamicin and infectivity of the variants in experimental animals. *J Infect Dis* 136(3), 360-9.

Neut, D., Hendriks, J. G., van Horn, J. R., van der Mei, H. C., and Busscher, H. J. (2005). Pseudomonas aeruginosa biofilm formation and slime excretion on antibiotic-loaded bone cement. *Acta Orthop* 76(1), 109-14.

Pan, X. S., Hamlyn, P. J., Talens-Visconti, R., Alovero, F. L., Manzo, R. H., and Fisher, L. M. (2002). Small-colony mutants of Staphylococcus aureus allow selection of gyrase-mediated resistance to dual-target fluoroquinolones. *Antimicrob Agents Chemother* 46(8), 2498-506.

Pelletier, L. L., Jr., Richardson, M., and Feist, M. (1979). Virulent gentamicin-induced small colony variants of Staphylococcus aureus. *J Lab Clin Med* 94(2), 324-34.

Proctor, R. (2000). "Respiration and small-colony variants of Staphylococcus aureus." Gram-positive pathogens (R. P. N. V. A. Fischetti, J. J. Ferretti, D. A. Portnoy, and J. I. Rood Ed.) ASM Press, Washington, D.C.

Proctor, R. A., Kahl, B., von Eiff, C., Vaudaux, P. E., Lew, D. P., and Peters, G. (1998). Staphylococcal small colony variants have novel mechanisms for antibiotic resistance. *Clin Infect Dis* 27 Suppl 1, S68-74.

Proctor, R. A., van Langevelde, P., Kristjansson, M., Maslow, J. N., and Arbeit, R. D. (1995). Persistent and relapsing infections associated with small-colony variants of Staphylococcus aureus. *Clin Infect Dis* 20(1), 95-102.

Proctor, R. A., von Eiff, C., Kahl, B. C., Becker, K., McNamara, P., Herrmann, M., and Peters, G. (2006). Small colony variants: a pathogenic form of bacteria that facilitates persistent and recurrent infections. *Nat Rev Microbiol* 4(4), 295-305.

Roggenkamp, A., Sing, A., Hornef, M., Brunner, U., Autenrieth, I. B., and Heesemann, J. (1998). Chronic prosthetic hip infection caused by a small-colony variant of Escherichia coli. *J Clin Microbiol* 36(9), 2530-4.

Rolauffs, B., Bernhardt, T. M., von Eiff, C., Hart, M. L., and Bettin, D. (2002). Osteopetrosis, femoral fracture, and chronic osteomyelitis caused by Staphylococcus aureus small colony variants (SCV) treated by girdlestone resection--6-year follow-up. *Arch Orthop Trauma Surg* 122(9-10), 547-50.

Rusthoven, J. J., Davies, T. A., and Lerner, S. A. (1979). Clinical isolation and characterization of aminoglycoside-resistant small colony variants of Enterobacter aerogenes. *Am J Med* 67(4), 702-6.

Sasarman, A., Surdeanu, M., Szegli, G., Horodniceanu, T., Greceanu, V., and Dumitrescu, A. (1968). Hemin-deficient mutants of Escherichia coli K-12. *J Bacteriol* 96(2), 570-2.

Seaman, P. F., Ochs, D., and Day, M. J. (2007). Small-colony variants: a novel mechanism for triclosan resistance in methicillin-resistant Staphylococcus aureus. *J Antimicrob Chemother* 59(1), 43-50.

Seggewiss, J., Becker, K., Kotte, O., Eisenacher, M., Yazdi, M. R., Fischer, A., McNamara, P., Al Laham, N., Proctor, R., Peters, G., Heinemann, M., and von Eiff, C. (2006). Reporter metabolite analysis of transcriptional profiles of a Staphylococcus aureus strain with normal phenotype and its isogenic hemB mutant displaying the small-colony-variant phenotype. *J Bacteriol* 188(22), 7765-77.

Seifert, H., Oltmanns, D., Becker, K., Wisplinghoff, H., and von Eiff, C. (2005). Staphylococcus lugdunensis pacemaker-related infection. *Emerg Infect Dis* 11(8), 1283-6.

Seifert, H., von Eiff, C., and Fatkenheuer, G. (1999). Fatal case due to methicillin-resistant Staphylococcus aureus small colony variants in an AIDS patient. *Emerg Infect Dis* 5(3), 450-3.

Seifert, H., Wisplinghoff, H., Schnabel, P., and von Eiff, C. (2003). Small colony variants of Staphylococcus aureus and pacemaker-related infection. *Emerg Infect Dis* 9(10), 1316-8.

Sendi, P., and Proctor, R. A. (2009). Staphylococcus aureus as an intracellular pathogen: the role of small colony variants. *Trends Microbiol* 17(2), 54-8.

Sendi, P., Rohrbach, M., Graber, P., Frei, R., Ochsner, P. E., and Zimmerli, W. (2006). Staphylococcus aureus small colony variants in prosthetic joint infection. *Clin Infect Dis* 43(8), 961-7.

Sherris, J. C. (1952). Two small colony variants of Staph. aureus isolated in pure culture from closed infected lesions and their carbon dioxide requirements. *J Clin Pathol* 5(4), 354-5.

Shih, H. N., Shih, L. Y., and Wong, Y. C. (2005). Diagnosis and treatment of subacute osteomyelitis. *J Trauma* 58(1), 83-7.

Sifri, C. D., Baresch-Bernal, A., Calderwood, S. B., and von Eiff, C. (2006). Virulence of Staphylococcus aureus small colony variants in the Caenorhabditis elegans infection model. *Infect Immun* 74(2), 1091-6.

Singh, R., Ray, P., Das, A., and Sharma, M. Enhanced production of exopolysaccharide matrix and biofilm by a menadione-auxotrophic Staphylococcus aureus small-colony variant (SCV). *J Med Microbiol.*

Soren, L., and Nilsson, L. (1984). Regrowth of aminoglycoside-resistant variants and its possible implication for determination of MICs. *Antimicrob Agents Chemother* 26(4), 501-6.

Swingle, E. L. (1935). Studies on Small Colony Variants of Staphylococcus aureus. *J Bacteriol* 29(5), 467-89.

Tappe, D., Claus, H., Kern, J., Marzinzig, A., Frosch, M., and Abele-Horn, M. (2006). First case of febrile bacteremia due to a wild type and small-colony variant of Escherichia coli. *Eur J Clin Microbiol Infect Dis* 25(1), 31-4.

Tsuji, B. T., von Eiff, C., Kelchlin, P. A., Forrest, A., and Smith, P. F. (2008). Attenuated vancomycin bactericidal activity against Staphylococcus aureus hemB mutants expressing the small-colony-variant phenotype. *Antimicrob Agents Chemother* 52(4), 1533-7.

Vann, J. M., and Proctor, R. A. (1988). Cytotoxic effects of ingested Staphylococcus aureus on bovine endothelial cells: role of S. aureus alpha-hemolysin. *Microb Pathog* 4(6), 443-53.

Vaudaux, P., Francois, P., Bisognano, C., Kelley, W. L., Lew, D. P., Schrenzel, J., Proctor, R. A., McNamara, P. J., Peters, G., and Von Eiff, C. (2002). Increased expression of clumping factor and fibronectin-binding proteins by hemB mutants of Staphylococcus aureus expressing small colony variant phenotypes. *Infect Immun* 70(10), 5428-37.

Vesga O, Groeschel MC, Otten MF, et al.: 1996, Staphylococcus aureus small colony variants are induced by the endothelial cell intracellular milieu. J Infect Dis 173:739-742.

von Eiff, C. (2008). Staphylococcus aureus small colony variants: a challenge to microbiologists and clinicians. *Int J Antimicrob Agents* 31(6), 507-10.

von Eiff, C., and Becker, K. (2007). Small-colony variants (SCVs) of staphylococci: a role in foreign body-associated infections. *Int J Artif Organs* 30(9), 778-85.

von Eiff, C., Becker, K., Metze, D., Lubritz, G., Hockmann, J., Schwarz, T., and Peters, G. (2001). Intracellular persistence of Staphylococcus aureus small-colony variants within keratinocytes: a cause for antibiotic treatment failure in a patient with darier's disease. *Clin Infect Dis* 32(11), 1643-7.

von Eiff, C., Bettin, D., Proctor, R. A., Rolauffs, B., Lindner, N., Winkelmann, W., and Peters, G. (1997a). Recovery of small colony variants of Staphylococcus aureus following gentamicin bead placement for osteomyelitis. *Clin Infect Dis* 25(5), 1250-1.

von Eiff, C., Heilmann, C., Proctor, R. A., Woltz, C., Peters, G., and Gotz, F. (1997b). A site-directed Staphylococcus aureus hemB mutant is a small-colony variant which persists intracellularly. *J Bacteriol* 179(15), 4706-12.

von Eiff, C., Jansen, B., Kohnen, W., and Becker, K. (2005). Infections associated with medical devices: pathogenesis, management and prophylaxis. *Drugs* 65(2), 179-214.

von Eiff, C., Vaudaux, P., Kahl, B. C., Lew, D., Emler, S., Schmidt, A., Peters, G., and Proctor, R. A. (1999). Bloodstream infections caused by small-colony variants of coagulase-negative staphylococci following pacemaker implantation. *Clin Infect Dis* 29(4), 932-4.

von Gotz, F., Haussler, S., Jordan, D., Saravanamuthu, S. S., Wehmhoner, D., Strussmann, A., Lauber, J., Attree, I., Buer, J., Tummler, B., and Steinmetz, I. (2004). Expression analysis of a highly adherent and cytotoxic small colony variant of Pseudomonas aeruginosa isolated from a lung of a patient with cystic fibrosis. *J Bacteriol* 186(12), 3837-47.

Wellinghausen, N., Chatterjee, I., Berger, A., Niederfuehr, A., Proctor, R. A., and Kahl, B. C. (2009). Characterization of clinical Enterococcus faecalis small-colony variants. *J Clin Microbiol* 47(9), 2802-11.

Wilson, S. G., and Sanders, C. C. (1976). Selection and characterization of strains of Staphylococcus aureus displaying unusual resistance to aminoglycosides. *Antimicrob Agents Chemother* 10(3), 519-25.

Wise, R. I., and Spink, W. W. (1954). The influence of antibiotics on the origin of small colonies (G variants) of Micrococcus pyogenes var. aureus. *J Clin Invest* 33(12), 1611-22.

Wu, M., von Eiff, C., Al Laham, N., and Tsuji, B. T. (2009). Vancomycin and daptomycin pharmacodynamics differ against a site-directed Staphylococcus epidermidis mutant displaying the small-colony-variant phenotype. *Antimicrob Agents Chemother* 53(9), 3992-5.

Youmans, G. P., and Delves, E. (1942). The Effect of Inorganic Salts on the Production of Small Colony Variants by Staphylococcus Aureus. *J Bacteriol* 44(1), 127-36.

CHAPTER IV

RESUSCITATION OF VIABLE BUT NON-CULTURABLE BACTERIA, OUTBREAK, AND THE GLOBAL SPREAD OF CHOLERA

Section 1
Viable but Non-Culturable Bacteria:
A Review

Traditionally, the viability of bacteria is related to its ability to form colonies on solid medium. When they lose the ability to form colonies, bacteria are considered dead. The usual laboratory procedure to determine the number of viable cells is the plate count method. In 1982, the laboratory of Rita Colwell, University of Maryland, reported that both *E. coli* and *V. cholerae* grown in salt water microcosms for 2 weeks remained viable even though they lost the ability to form visible colonies on agar (Xu *et al.* 1982). This state, namely the viable but non-culturable (VBNC) state, is considered to be a survival strategy of bacteria against starvation and stress. Thus under unfavorable conditions, bacteria may enter the VBNC state and exhibit very low levels of metabolic activity and remain unculturable. However, upon removal of stress and addition of nutrients, VBNC cells may resuscitate and then grow normally and become culturable.

The ability of bacteria to enter the VBNC state was further supported by the findings that *V. cholerae* could remain in water for a long time in the VBNC state when the temperature of water dropped below 10°C (Huq *et al.* 1990; Colwell *et al.* 1996; Colwell 1996; Chaiyanan *et al.* 2001). Thus the difficulty in isolating bacteria from the water samples during the winter season was explained on the basis of the ability of bacteria to enter the VBNC state (Huq *et al.* 1990; Colwell 1996). During the summer, when the temperature increases, they may exit the VBNC state, become culturable and multiply, causing fresh rounds of infection. Whitesides and Oliver (1997) reported a drastic drop in the colony counts when *Vibrio vulnificus* were grown in artificial sea water (ASW) at 5°C. After 4-5 days, the number of colonies was reduced to below 10 cfu/ml. At this stage, bacteria were considered to enter the VBNC state. However, direct examination of cells in the VBNC state indicated that those cells were intact and had not lysed.

As cells enter the VBNC state, a decrease in DNA, RNA and protein synthesis occurs in most cases (Lebaron and Joux 1994; Effendi and Austin 1995; Weichart *et al.* 1997). However, protein synthesis is not completely stopped but rather a change in the protein profile occurs, with the production of many new proteins not observable during the growth of bacteria at normal temperatures (Effendi and Austin 1995). However, proteins specific to the VBNC state have not been detected (Meula *et al.* 2008). Even though a reduction in the nucleic acid content occurs as the cells enter the VBNC state, some VBNC cells may maintain normal amounts of DNA (Mukamolova *et al.* 1995; Weichart *et al.* 1997). *Micrococcus luteus* VBNC cells, for example, maintained normal DNA amounts even though the RNA amount was reduced to 50% (Mukamolova *et al.* 1995). In spite of the reduction in nucleic acid and protein synthesis, VBNC cells may maintain high levels of ATP (Beumer *et al.* 1992), continue gene expression (Lleo *et al.* 2000), actively take up methionine to incorporate into proteins (Rahman *et al.* 1994) and maintain membrane fluidity by changing the membrane fatty acid composition (Day and Oliver 2004). When rod-shaped bacteria become nonculturable, their size reduces and their shape changes from rod to coccus (Nilsson *et al.* 1991; Oliver *et al.* 1991; Oliver 2005; Chaiyanan *et al.* 2007). This size reduction and change in morphology to the coccoid forms is considered to be a protective strategy against adverse conditions (Chaiyanan *et al.* 2007).

Different methods have been used to determine the viability of bacteria apart from the usual plate methods. These include direct viability counts, tetrazolium salt reduction (del Campo *et al.* 2009), acridine orange direct counting (Heidelberg *et al.* 1997), use of enzyme substrates such as fluorescein diacetate (Rahman *et al.* 2001), multiplex PCR assay (Vora *et al.* 2005), mRNA detection using reverse transcription-PCR (Lleo *et al.* 2000), nucleic acid staining etc. (Weichart *et al.* 1997). Each method has its own strengths and weaknesses, and any single method may not be sufficient to correctly detect the VBNC state (Rice *et al.* 2000). For example, when Ethidium bromide monoazide real-time PCR (EMA Rti-PCR) amplification of target DNA was compared with the plate count assay for calculating the number of viable cells induced by exposing *Vibrio vulnificus* at -20 and 4°C, the log CFU values from the EMA Rti-PCR assays were much higher than that of the plate counts. This might indicate that a considerable number of viable cells were in the VBNC state. However, when a combination of sodium deoxycholate (SD) and EMA treatments were used, a high correlation was noticed between the plate counts and the number of viable cells determined from SD+EMA Rti-PCR (Lee and Levin 2009). Thus the use of EMA alone may give a large number of 'false positive' viable cells.

Similarly, other labeling methods that use indirect, non-culture-based methods can give 'false positive' viable cell numbers (Bogosian and Bourneuf 2001).

Transmission electron micrographs of *Campylobacter jejuni* VBNC cells revealed an intact but asymmetric formation of membranes, condensed cytosol (Rollins and Colwell 1986) and reduced ribosomal and nucleic acid densities (Chaiyanan *et al.* 2001; Linder and Oliver 1989). Thus VBNC cells may maintain an intact cell membrane to keep the genetic material intact, but at the same time may maintain low metabolic activity to help their survival during starvation and other adverse conditions (Linder and Oliver 1989). Differences in the morphological shape of *V. cholerae* VBNC cells were noticed when the cells were grown at room temperature and at 4°C (Chaiyanan *et al.* 2007). For those bacteria grown at room temperature, the morphology changed over time; they were initially rods with a single flagellum, but gradually changed to irregular shaped rods with or without flagella and later to coccoid cells (Chaiyanan *et al.* 2007). When the temperature was raised, the coccoid cells reverted to rod-shaped cells with flagella and underwent active division. On the other hand, when *V. cholerae* were grown at 4°C, they entered the VBNC state within 2 months (Chaiyanan *et al.* 2007). At the end of the first month, irregular rods and coccoids were observed. When heat-treated at 45°C for 1 min, only the irregular rods, but not the coccoids, were capable of growth on LB agar. However, coccoid cells were also viable as confirmed by the tetrazolium salt reduction method. Thus starvation and low temperatures may show different stress responses (Chaiyanan *et al.* 2007). This was also supported by the proteome analysis of starved and VBNC cells of *Enterococcus feacalis* (Heim *et al.* 2002). During starvation, the cells underwent reduced metabolism, but did not lose their culturability (Oliver *et al.* 1991). On the other hand, cells entered the VBNC state and lost their culturability at low temperatures.

A number of factors can induce bacteria to enter the VBNC state (Oliver 2005). This includes changes in temperature, starvation, UV light, osmotic stress, stationary phase of growth, chemicals etc. (Oliver 2005; Trainor *et al.* 1999; Wood *et al.* 2005). Whereas some bacteria enter the VBNC state in a few hours or days (Klancnik *et al.* 2009), others require several days or even months (van Overbeek *et al.* 2004) to enter the VBNC state.

Initially, the ability of bacteria to enter the VBNC state was reported in *Vibrio* and *E. coli*, but later, it was reported in a number of bacteria (both gram-negative and gram-positive) including human pathogens like *Camphylobacter, Listeria, Helicobacter, Pseudomonas, Mycobacterium, Legionella* and *Salmonella* (Oliver 2005). The important grampositive organisms capable of entry into the VBNC state include *Listeria, Enterococcus, Streptococcus* (Trainor *et al.* 1999; Wood *et al.* 2005) and *Micrococcus* (Mukamolova *et al.* 1995). Another gram-positive human pathogen, *Staphylococcus aureus,* may not enter the VBNC state under starvation (Diaper and Edwards 1994; Watson *et al.* 1998). When *S. aureus* were grown in nutrient limited conditions, 99-99.9% of bacteria lost their culturability within 2 days. Although the surviving cells remained viable for months, there was no evidence of VBNC formation since the number of bacteria determined by CFU counts was not significantly different from the viable counts (Diaper and Edwards 1994; Watson *et al.* 1998).

Because of the inadequacies in detecting viable bacteria using standard procedures, VBNC and its resuscitation is important to public health from the standpoint of food and water safety (Oliver 2005). Researchers supporting the VBNC state maintain that it is a bacterial survival strategy (Roszak and Colwell 1987) and a genetically programmed response to adverse conditions. Resuscitation to normal growth during favorable conditions may help in maintaining the population. Meula *et al.* (2008) consider VBNC to be, rather than a successful phenotype, an altruistic strategy to maintain the bacterial population. They argue that VBNC cells may not recover culturability but may excrete organic molecules that can be used by the few remaining culturable bacteria under unfavorable conditions.

Resuscitation of VBNC cells of various bacteria has been demonstrated *in vitro, in vivo* and *in situ* (Oliver 2005). For some bacteria, resuscitation requires the addition of nutrients, whereas for others, the removal of the stress condition that induced the VBNC state is sufficient for resuscitation. For example, VBNC *V. vulnificus* induced by low temperature could be resuscitated by a simple upshift of temperature (Oliver *et al.* 1991; Oliver and Bockian 1995). How an increase in the temperature helped in resuscitation was not known, but the authors suggested that certain factors might be needed for the resuscitation of VBNCs. Later, a group of researchers found that a low molecular protein, namely resuscitation promoting factor (Rpf) produced by the growing cells of *Micrococcus luteus* (Mukamolova *et al.* 1995; Telkov *et al.* 2006), was needed for the resuscitation of dormant cells of the same organism. Addition of Rpf, released by the late log phase of viable cultures, could resuscitate the dormant cells. Rpf is a peptidoglycan hydrolyzing enzyme, and resuscitation requires the hydrolysis of peptidoglycans (Telkov *et al.* 2006). Rpf proteins are also produced by *Mycobacterium tuberculosis* (Mukamolova *et al.* 2010) and may play an important role in the *in vivo* persistence of tuberculous bacteria. Similarly, an Rpf-like protein has been described in the *Salmonella typhimurium* strain LT2 also (Panutdaporn *et al.* 2006).

One of the problems encountered with in resuscitation experiments is the difficulty of proving that the growth seen was due to 'true resuscitation' of VBNC and not due to the regrowth of a few culturable bacteria. Whitesides and Oliver (1997) diluted a VBNC population of *V. vulnificus* to one thousand-fold so that samples contained as few as 10^3 VBNC cells and less than 0.0001 culturable bacteria/ml. Bacterial growth was still noticed after incubation of these cells at room temperature, indicating 'true resuscitation' of VBNC and not the regrowth of a few culturable bacteria.

Resuscitation of VBNCs and their significance have faced severe criticism as well (Barer 1997; Bogosian *et al.* 2000; Bogosian and Bourneuff 2001; Nystrom 2001). Many researchers contend that the terminology itself is wrong as 'true VBNCs' cannot be resuscitated under any circumstances (Barer 1997). Others suggest that the growth recovery is due to the regrowth of a few culturable bacteria only and not due to the resuscitation of VBNCs (Bogosian *et al.* 1998). Growth recovery can also be due to the re-growth of a subpopulation of injured cells (Bogosian and Bourneuff 2000). When bacteria are exposed to stress conditions, a subpopulation may become injured and, even though they may maintain viability, they may not be able to form colonies (Bogosian and Bourneuff 2000). Upon the removal of stress conditions, some of the injured cells may undergo repair and start to regrow. Thus, the critics of VBNC argue that the VBNC state is not a genetically programmed phenomenon. Bogosian *et al.* (2000) suggested that hydrogen peroxide sensitive culturable cells could be the basis of the resuscitation of VBNC. They noticed that the culturability of *V. vulnificus* grown at 5°C was lost completely within 2-3 weeks, even though the total cell counts were unchanged. However, when the agar plates contained catalase or sodium pyruvate, re-growth of culturable bacteria occurred. They suggested that the presence of hydrogen peroxide in agar plates inhibited the growth of a normal subpopulation of H_2O_2-sensitive cells and that the removal of H_2O_2 by catalase resulted in the re-growth of that subpopulation giving the appearance of resuscitation of VBNC. During the gradual loss of culturability, cells may pass through a transient H_2O_2-sensitive injured state before progressing to complete loss of culturability.

Thus, the most important question pertaining to VBNCs is whether they have the ability to resuscitate to normal growth on the removal of the stress that forced them to enter the VBNC state and/or by the addition of nutrients. In the next section, reviews of the literature that supports and opposes the conclusion of resuscitation and the significance of VBNC cells are given under different headings. Each of the findings is explained for clarity rather than just citing the references.

A. Findings that support the resuscitation of VBNC

1. *Vibrio cholerae* can be detected in the aquatic environment throughout the year (Kaper *et al.* 1979; Alam *et al.* 2006a; Alam *et al.* 2006b; Louis *et al.* 2003; Binsztein *et al.* 2004), even though it is difficult to isolate culturable bacteria by conventional culture methods during the inter-epidemic periods (Huq *et al.* 1990). During the winter, when the conditions are unfavorable for growth, bacteria may remain in the VBNC state, but may resuscitate to normal virulent forms once the temperature and other conditions become favorable for growth.

2. *V. cholerae* maintained in a laboratory microcosm entered the VBNC state and changed from rod to coccoid morphology. Those VBNC cells retained their viability upto one year and maintained pathogenicity as well as chromosomal integrity (Chaiyanan *et al.* 2001)

3. When VBNC *Campylobacter jejuni*, induced by growing bacteria in surface water for 30 days, were inoculated into the yolk sacs of embryonated eggs, culturable cells were detected in a large proportion in the embryonated eggs, which also maintained their adhesion properties (Cappelier *et al.* 1999b).

4. VBNC cells from three *C. jejuni* human isolates, obtained by growing in microcosm water at 4°C, were inoculated *per os* into newborn mice and 1-day-old chicks. All three strains were revived in the murine model, whereas two strains were revived in the chick model (Cappelier *et al.* 1999a)

5. When 10^4 metabolically active VBNC cells of *Listeria monocytogenes* were inoculated into the vitellus fluid of embryonated and non-embryonated eggs, culturable bacteria were detected in many of the embryonated eggs (Cappelier *et al.* 2007).

6. *Erwinia amylovora* adopted a VBNC state when grown at 26°C or when treated with copper sulfate. VBNC cells, maintained for at least 28 days, recovered culturability in KB broth and maintained pathogenicity after host plant passage (Ordax *et al.* 2009).

7. Passage of VBNC *C. jejuni*, induced by growth in bottled water at 4°C in the dark, into embryonated eggs resulted in the recovery of cells following resuscitation of VBNC cells (Guillou *et al.* 2008).

8. *Escherichia coli* O157H-strain VBNC, grown in sterilized distilled water at 4°C, were resuscitated by growing in a solid medium containing catalase, sodium pyruvate or α-ketoglutaric acid indicating that cells enter VBNC on oxidative stress and that the correction of stress helps in the resuscitation of VBNC (Mizunoe *et al.* 1999).

9. *Legionella pneumophila* entered the VBNC state after growing in sterile tap water for 125 days. Addition of the natural host, *Acanthamoeba castellanii*, to the non-culturable bacteria resulted in the uptake of VBNC cells by the host followed by their intracellular multiplication and resuscitation to a culturable state. However, resuscitation was not observed in the guinea pig model (Steinert *et al.* 1997)

10. During the winter, when the temperature of natural estuarine water was below 15°C, *V. vulnificus* entered the VBNC state. However, those VBNC cells underwent rapid resuscitation to culturable forms during the warmer months when the temperature was above 21°C indicating that a simple temperature upshift is sufficient to resuscitate VBNC cells (Oliver *et al.* 1995)

11. *Ralstonia solanacaerum* entered the VBNC state in liquid microcosms, sterile soil and *in planta*. VBNC cells in sterile soil resuscitated and infected germinated tomato seeds (Grey and Steck 2001a)

12. Enterotoxigenic *E. coli* (ETEC) can survive in fresh water and seawater for long periods and can enter the VBNC state. Though enterotoxins were not produced in the VBNC state, they expressed genes coding for ETEC toxins, colonization factors and metabolic pathways (Lothigius *et al.* 2010).

13. After monochloramine treatment at 1 mg/L, *Legionella pneumophila* entered VBNC but regained its culturability when cultivated in the presence of the amoeba *Acanthamoeba castellanii* (Alleron *et al.* 2008). Similarly, Dusserre *et al.* (2008) reported that VBNC *L. pneumophila*, induced by chlorinated water, could be resuscitated after 5 days of

coculture with *A. polyphaga*. Similar results were reported by Hwang *et al.* (2006).

14. Strains of *Vibrio alginolyticus* and *V. parahaemolyticus* VBNCs, produced in ASW microcosms at 5°C, were resuscitated in a murine model by intragastric inoculation of approximately 10^7 organisms. The resuscitated organisms were less virulent but could be reactivated after two consecutive passages in the rabbit ileal loop model (Baffone *et al.* 2003)

15. *V. cholerae* O1 cells lost their culturability after incubation in ASW at 5°C or 18°C, but retained the ability to adhere to chitin particles, copepods and human intestinal epithelial cells, although with less efficiency (Pruzzo *et al.* 2003). Similarly, *Enterococcus faecalis* VBNCs maintained their ability to adhere to Caco-2, Girardi heart cultured cells and urinary tract epithelial cells, but with 50-70% less efficiency (Pruzzo *et al.* 2002). The results were consistent with the findings of Signoretto *et al.* (2004) and Signoretto *et al.* (2005), who also reported that *E. faecalis* VBNC adhered to copepods and chitin, though less efficiently than to growing cells.

16. *Streptococcus parauberis* entered the VBNC state when cultured in seawater and sediment microcosms. The addition of fresh medium to microcosms resuscitated VBNCs, which maintained their virulence in turbots resulting in the death of fish (Curras *et al.* 2002)

17. Camphylobacter strains under acidic conditions (formic acid at pH 4) lost their culturability, but not viability, after 2 h. However, culturability was restored via passage through both the allantoic and yolk sac routes of embryonated eggs (Chaveerach *et al.* 2003).

18. VBNC *V. vulnificus* could be resuscitated by temperature upshift. Using dilution methods and determining the time taken for the appearance of culturable bacteria, it was found that the recovery of growth was due to true resuscitation and not due to the regrowth of a few culturable bacteria (Whitesides and Oliver 1997).

19. *V. vulnificus* VBNC, induced at 5°C, when injected into an iron overload mouse model, were found to be virulent and caused the death of the mouse. Culturable *V. vulnificus* were recovered from the blood and peritoneal cavities of mice (Oliver and Bockian 1995).

20. Temperature upshift to 56°C for 15 s followed by incubation at 37°C onto TSA plates resuscitated *Salmonella enterica serovar Typhimurium DT104* VBNC maintained at 5°C for 250-300 days (Gupte *et al.* 2003)

21. *V. vulnificus* biotype 2 VBNC could maintain their infectivity in eels and mice (Marco-Noales *et al.* 1999)

22. A starved population of *Micrococcus luteus* lost its culturability in 3-6 months, but was resuscitated by incubating in liquid medium containing the supernatant taken from the late log phase of viable cultures of the same organism (Mukamolova *et al.* 1995). The supernatant contains a protein named 'resuscitation promoting factor', a peptidoglycan-hydrolyzing enzyme (Telkov *et al.* 2006).

23. VBNC *C. jejuni* strains, maintained at 4°C in ASW, survived for 138 to 152 days. When the number of respiring bacteria in the VBNC state was more than 10^4 cells/ml, resuscitation was noticed following their passage into the mouse intestine (Baffone *et al.* 2006).

B. Findings that reject the resuscitation of VBNC

1. VBNC *L. monocytogenes*, maintained in microcosm water at 4°C could not be resuscitated, nor were they virulent in a cell plaque assay or by intraperitoneal inoculation into immunodeficient mice (Lindback *et al.* 2010).

2. VBNC *C. jejuni* suspended in water for 7-14 days were given to day-of-hatch chickens and their caeca were examined for the presence of bacteria after one or two weeks of challenge. No culturable bacteria were detected in the caeca, thus questioning the ability of VBNC to revert to culturable form (Ziprin *et al.* 2003; Ziprin and Harvey 2004).

3. When culturable bacteria become VBNC, rod-shaped bacteria change to coccoidal forms. Introduction of coccoidal VBNC *C. jejuni* into stimulated gastric, ileal and colon environments did not revert VBNCs to culturable forms nor did they cause any infection in laboratory animals or volunteers after oral administration (Beumer *et al.* 1992).

4. Coccoid forms of *Prolinoborus fasciculus* exhibited extensive rRNA degradation and could not be resuscitated, indicating that they were only degenerative forms of bacteria (Koechlein and Krieg 1998). Similarly, Boucher *et al.* (1994) found that the coccoidal forms of *C. jejuni* could not be sustained in adverse conditions for prolonged periods and concluded that they were degenerative forms rather than bacteria in a dormant state.

5. Under low relative humidity, VBNC *L. monocytogenes* survived on parsley leaves but did not revert to culturable forms following trans-

fer to high humidity. Only residual culturable forms, and not VBNC, were able to grow under increased humidity (Dreux *et al.* 2007).

6. VBNC *C. jejuni* introduced to day-old chicks or passage through allantoic fluid of embryonated eggs could not be resuscitated, and no organism could be isolated from the caeca of chicks after 7 days of incubation, indicating that they lacked the ability to colonize the intestine (Medema *et al.* 1992).

7. VBNC *S. enterica serovar Typhimurium*, rendered non-culturable by carbon and nitrogen stress, could not colonize any organs, nor could they cause infections when administered orally or intraperitoneally to more than 300 female BALB/c mice (Smith *et al.* 2002).

8. Following VBNC state induction of *E. coli* O157:H7 by growing in sterile water or after exposure to chlorine, culturable bacteria could not be recovered *in vitro* even in the presence of catalase or sodium pyruvate in the culture medium. Similarly, the passage of VBNC cells through the gastrointestinal tract by challenging mice orally with VBNC cells did not help in the recovery of culturable bacteria. They were unable to cause cytotoxicity in mouse kidney samples suggesting lack of virulence of VBNC cells (Kolling and Mathews 2001).

9. When INT-407 cells were exposed to culturable *C. jejuni* with or without VBNC cells of *C. jejuni*, no differences were found in the number of bacteria invading or adhering to INT-407 cells, and the levels of IL-8 secretion indicating that VBNC cells lack virulence and are unable to induce any immune response (Verhoeff-Bakkenes *et al.* 2008).

10. In the *E. coli* K-12 strain, the VBNC state was induced by growing bacteria in sterile and non-sterile water and soil at different temperatures. While VBNC cells in sterile water or soil persisted for more than 50 days without much decline in their numbers, those in non-sterile water or soil were reduced by an order of log 3 to 5. Non-culturable cells in either sterile or non-sterile water or soil could not be resuscitated, indicating that VBNC cells are non-viable (Bogosian *et al.* 1996).

11. *E. coli* cells exposed to many adverse conditions were rendered non-culturable. Resuscitation of those VBNC cells was attempted by removing the stress factors that induced the VBNC state, adding nutrients and exposing the cells to lysozyme and culture supernatants. However, no resuscitation could be detected with any of the methods. Moreover, the VBNC subpopulation was not the predominant

population among the culturable, VBNC and non-viable subpopulations, indicating that VBNC is not a successful phenotype (Arana *et al.* 2007).

12. VBNC *L. monocytogenes* produced by microcosm water model could not be resuscitated by normal resuscitation methods. They were avirulent and could not adhere to or invade HT-29 cells, nor could they resuscitate and cause infection in the mouse model (Cappelier *et al.* 2005).

13. Chlorination of secondary treated waste water containing *E. coli* and *S. typhimurium* resulted in the loss of culturability but not viability. Those cells could not be resuscitated by temperature upshift or nutrient addition (Oliver *et al.* 2005). However, the authors maintained that VBNC could be a potential public health hazard in spite of its incapability of resuscitating.

14. The ability of *C. jejuni* grown in 0.9% saline at 2-4°C to colonize day-old chicks was tested. The ability to colonize was reduced during the incubation period and terminated 3-4 weeks before the loss of culturability, thus questioning the importance of VBNC in initiating infections (Hald *et al.* 2001).

15. *S. typhimurium* VBNCs maintained in ASW and exposed to UV-C were tested for pathogenicity and virulence in the mouse model. VBNC cells lacked the pathogenicity to kill the mouse even though they were viable (Caro *et al.* 1999).

16. The inability to detect *V. vulnificus* VBNC cells during winter from the water samples collected from Barnegat Bay, N.J., using MPN-QPCR assay, coupled with the inability to resuscitate VBNC cells, suggest that during winter, *V. vulnificus* disappear from the water column and that the appearance of the organism during summer may not be due to resuscitation of VBNC (Randa *et al.* 2004).

17. VBNC *Camphylobacter* spp., formed after growing bacteria in sterilized water and potassium phosphate buffer at 4°C, could not be recovered from mice or chicks after 4 and 2 weeks, respectively (van de Geissen *et al.* 1996).

18. Culturability of *C. jejuni* and adhesion/invasion was found to be linear; thus the absence of culturable bacteria during the VBNC state can be considered to be a low risk, at least in the case of camphylobacteriosis (Verhoeff-Bakkenes *et al.* 2009).

19. VBNC cells of *Francisella tularensis* could not be resuscitated, nor could they cause tularemia in mice, when 10^5 VBNC cells were injected intraperitoneally (Forsman *et al.* 2000).

20. In a river microbial community, *E. coli* cells are ingested by flagellates or other predators even before cellular deterioration occurs. Thus predation reduces the importance of VBNC cells in natural aquatic systems (Arana *et al.* 2003).

C. *In vitro*, but not *in vivo*, resuscitation of VBNC cells

1. VBNC cells of two clinical strains of *Vibrio harvey* were resuscitated *in vitro* by temperature uplift and in the presence of Tween 20 or vitamin B. Zebra fish inoculated with resuscitated cells died in a week, but no death was noticed in the group inoculated with VBNC cells (Sun *et al.* 2008). In another similar experiment, VBNC *V. cincinnatiensis*, cultured in sterilized seawater microcosms at 4°C, were resuscitated by temperature upshift in media containing yeast extract and peptone along with catalase or vitamin B. Here also, zebra fish inoculated intraperitoneally with resuscitated cells died, but no death was noticed in the group inoculated with VBNC cells (Zhong *et al.* 2009).

2. *Edwardsiella tarda*, a fish pathogen, entered the VBNC state after 28 days of incubation in the seawater microcosm at 4°C. Turbots inoculated with normal and resuscitated *E. tarda* died in 5-7 days, whereas those inoculated with VBNC cells and autoclaved saline survived during the experimental period, and no organism could be isolated from those animals (Du *et al.* 2007b).

3. *Vibrio alginolyticus* VIB283 VBNC cells cultured in a sterilized seawater microcosm at 4°C were resuscitated *in vitro* by temperature upshift with and without the presence of nutrition, but not in chick embryos (Du *et al.* 2007a).

4. After several days of incubation under anoxic conditions, *Sinorhizobium meliloti* entered the VBNC state and could be resuscitated at atmospheric O_2 supplementation. The *in vitro* resuscitated cells were able to nodulate *Medicago sativa*, whereas VBNCs were unable to do so (Basaglia *et al.* 2007).

5. *Aeromonas hydrophila* entered the VBNC state when cultured in 0.35% NaCl at 25°C for 50 days and could be resuscitated *in vitro* but were not pathogenic to goldfish (Rahman *et al.* 2001).

6. VBNC *Shigella dysenteriae* type 1 maintained its virulent factors and adhered to Henle 407 cells but lost its ability to invade the cells (Rahman *et al.* 1996) .

7. *V. cholerae* O139 VBNC cells maintained in ASW at 4°C could be resuscitated *in vitro*, but mice challenged with VBNC cells or resuscitated cells survived, whereas 80% of mice challenged with culturable cells died (Sung *et al.* 2006).

D. Growth recovery can be due to the regrowth of culturable bacteria and not due to the resuscitation of VBNC cells.

1. Bogosian *et al.* (1998) used a mixed culture recovery method to study whether growth recovery was due to resuscitation of VBNCs or due to the re-growth of a few culturable bacteria. In this method, a large number of nonculturable bacteria was mixed with a small number of culturable ones by using two easily distinguishable strains. On applying various resuscitation techniques, it was found that only the culturable bacteria could be recovered, indicating that growth recovery was due to a few culturable bacteria left and not due to the VBNC cells (Bogosian *et al.* 1998).

2. By dilution and temperature upshift experiments, Chowdhury *et al.* (1994) studied the resuscitation of VBNC cells of *V. cholerae* and *E. coli*. They noticed that the uptake of thymidine was below the detection level when the cells entered the VBNC state. Similarly, the uptake of glucose and acetate also decreased to very low levels when they entered the VBNC state. They found that the recovery on temperature upshift was indeed due to the regrowth of a few culturable cells and not due to the resuscitation of VBNC cells. When there were no culturable cells, no recovery was noticed (Chowdhury *et al.* 1994).

3. Culturable cells of *V. cholerae* were found to be more infectious than VBNC cells indicating that the major contributors of infection were culturable cells only. Moreover, 24 h of adaptation in pond water resulted in the loss of ability of *V. cholerae* to colonize the animals, thus challenging the importance of VBNC in causing cholera outbreaks (Nelson *et al.* 2008).

4. *V. cholerae* entered the VBNC state after growing in ASW at 4°C. Dilution and temperature upshift experiments proved that the recovery was due to the regrowth of a few culturable bacteria and not due to the true resuscitation of VBNC cells (Ravel *et al.* 1995).

5. A cold-incubated *V.vuinificus* population included culturable, non-culturable and injured populations. Regrowth was mainly due to culturable bacteria and a small proportion of the injured population but not due to the non-culturable bacteria, which could not be resuscitated by temperature upshift nor by the addition of nutrients or culture supernatants (Weichart and Kjelleberg 1996). In spite of the absence of recovery of VBNC under laboratory conditions, the authors concluded that the stability of VBNC cells might allow them to persist in the environment for long periods of time.

6. Resuscitation of *V. vulnificus* VBNC by temperature upshift was due to regrowth of a few culturable bacteria that utilized the substrates released from the dead ones (Weichart *et al.* 1992).

7. Resuscitation of *C. jejuni* from a stationary phase culture with a microaerobic gas mixture was studied. The most probable number (MPN) technique demonstrated that the multiplication of residual viable cells and a small proportion of bacteria in the injured or latent stage were responsible for the increase in counts (Bovill and Mackey 1997).

8. Cold temperatures and CO_2 killed the majority (69%) of *L. monocytogenes*, leaving 31% of stressed population that could include healthy, injured and putative VBNC cells. However, the selective plate count showed that the stressed population included only the healthy and injured cells but not the putative VBNC cells. Thus, the regrowth of bacteria after stress treatment could be due to the culturable bacteria only, which included both healthy and injured cells (Li *et al.* 2003).

9. Recovery of starved *V. parahaemolyticus* after temperature upshift was due to regrowth of a few surviving cells (Jiang and Chai 1996).

E. Only a small percentage of VBNC may resuscitate.

1. VBNC cells of *V. parahaemolyticus*, grown in ASW at 4°C, were resuscitated by a temperature upshift to 20°C or 37°C. Experiments indicated that only a portion of VBNCs were able to be resuscitated following the temperature upshift and that the regrowth of this small population was responsible for the growth recovery (Coutard *et al.* 2007).

2. When 10^6 mL^{-1} of *Xanthomonas axonopodis pv. citri* were treated with 135-μM $CuSO_4$ solution for 10 minutes, the bacteria lost their culturability. However, 16% of them were viable (VBNC), and only 1%

was virulent enough to produce cankers in grapefruit plants (del Campo et al. 2009). This suggests that not all VBNCs can be resuscitated to normal virulent form and that the resuscitation result from a small percentage of VBNCs.

3. Filtered and sterilized lake water was used to induce the VBNC state in three enterococcal strains and resuscitation was attempted by growing the VBNC cells on TSA plates or TSB medium at 37°C for 4 days. E. faecium were unculturable after 4 weeks, and no resuscitation was noticed after this period, indicating that they tend to die instead of remaining in the VBNC state. On the other hand, approximately 1 in 10^4 cells of VBNC E. faecalis and E. hirae could be resuscitated, but the number of resuscitable cells progressively declined over time (Lleo et al. 2001).

4. VBNC cells of V. vulnificus maintained cellular integrity and intact nucleic acids even after exposure to cold for long periods of time. However, on prolonged incubation, DNA and RNA degradation occurred and resulted in loss of viability. The majority of VBNC were viable, but only a small fraction of the subpopulation retained the ability to be resuscitated (Weichart et al. 1997).

F. VBNC can be sublethally injured cells or those nearing death.

1. C. jejuni entered the VBNC state after starvation for 5 h, survived in vitro within Caco-2 enterocites up to 4 days and was able to cause systemic infection in vivo in a mouse model. However, the number of bacteria was lower in many organs, and the mice quickly recovered from the infection. The authors concluded that the resuscitation of VBNC could be due to sublethally injured cells that had lost much of their virulence (Klancnik et al. 2009).

2. When V. parahaemolyticus entered VBNC after incubation at 4°C in ASW for 69 days, two groups of bacteria were detected: one with high nucleic acid (HNA) content and the other with low nucleic acid (LNA) content. The LNA population increased as the microcosm underwent aging, whereas the HNA population decreased. A good correlation was noticed between the LNA population and the dead cells, indicating that LNA could be either dead cells or those close to death (Falcioni et al. 2008).

3. Aeromonas hydrophila growth behavior in ASW at 5°C and 23°C indicated that they entered the VBNC state after 3-5 weeks of incubation. With the passage of time, there was successive loss of cul-

turability and membrane integrity, as well as an increase in the LNA subpopulation. On temperature upshift, recovery was noticed, which was mainly due to culturable bacteria and to some extent by damaged cells (Maalej *et al.* 2004).

4. *Ralstonia solanacearum biovar 2 strain 1609*, grown in water at 4°C, entered VBNC after 84 days. Virulence was found to decrease progressively. Until 84 days, when a few culturable bacteria were present, they were able to cause wilting of tomato plants. From 84-100 days, they were not able to cause wilting nor were they virulent, even though they maintained the cellular forms capable of proliferation. After 125 days, they were no longer able to proliferate, nor could they cause wilting, suggesting that VBNC cells progressively become injured before finally undergoing death (van Overbeek *et al.* 2004).

5. *Vibrio anguillarum* lost its culturability but not viability when exposed to sterilized lake water, but could not be resuscitated either *in vivo* by injecting into fish or *in vitro*. The authors concluded that the VBNC state is one that leads to death (Eguchi *et al.* 2000).

6. After one week of incubation at 4°C in the presence of CO_2, 69% of *L. monocytogenes* were found to be dead while 31% were live by LIVE/DEAD BacLight™ assay. The bacteria that were able to grow were found to be either normal healthy bacteria or injured ones but not VBNC (Li *et al.* 2003).

7. Exposure of *E. coli* to hypochlorous acid (HOCl) resulted in the appearance of 3 subpopulations: dead, live and VBNC. The recovery of the HOCl-stressed population in a phosphate buffer was found to be due to the growth of culturable bacteria and a few of the sublethally injured VBNC cells grown at the expense of damaged cells (Dukan *et al.* 1997).

8. Using *in situ* detection of protein oxidation in *E. coli* and density gradient centrifugation techniques to separate the culturable and non-culturable populations, Desnues *et al.* (2003) showed that the proteins in non-culturable cells underwent irreversible oxidative damage that resulted in the loss of culturability and deterioration of the cell. Thus VBNCs represent those bacterial subpopulations at the final stages of death.

9. Starvation of *E. coli* in river water at 20°C resulted in the reduction of the number of culturable cells, a reduction in chromosomal and

plasmid DNA content and an increase in 'ghost forms', indicating that starvation leads to gradual cellular death (Muela *et al.* 1999).

10. Non-culturable *Legionella pneumophila* obtained after 100 days of starvation showed chromosomal DNA and rRNA subunit degradation and could not be resuscitated in *Tetrahymena pyriformis*, one of its natural protozoan hosts. This indicates that VBNC cells probably have non functional nucleic acids and are unable to multiply (Yamamoto *et al.* 1996).

11. Copper induced *Xanthomonas campestris pathovar campestris* to the VBNC state within 2 days. As the concentration of cupric sulfate increased, the concentration of dead cells increased and that of VBNC decreased. Plating VBNC cells in a copper-free medium and adding chelators did not help to regain culturability even after 3 months (Ghezzi and Steck 1999). This indicates that VBNC cells are those nearing death. At low concentrations of copper, many of them may remain in an injured state, but at high concentrations, most of them become moribund.

12. During the aging of *Helicobacter pylori*, bacillary forms were converted to coccoid forms along with the gradual loss of culturability. Those non-culturable bacteria were only dead and degenerated forms and not VBNCs (Enroth *et al.* 1999).

G. Resuscitation depends on the length of time after remaining in the VBNC state.

1. VBNCs of *E. coli*, induced by copper, could be resuscitated for up to 2 weeks after entering the VBNC state (Grey and Steck 2001b). This indicates that the ability to resuscitate from the VBNC state is time-dependent and that they may lose their culturability completely after a period of time.

2. VBNCs of *Erwinia amylovora* induced by copper sulfate could maintain their virulence only for the first 5 days (Ordax *et al.* 2006).

3. *V. vulnificus biotype 2* VBNC cells, grown in ASW at 5°C, could be resuscitated by a temperature upshift alone. However, resuscitation failed when the length of time after entering the VBNC state was prolonged or when small sample sizes were used (Biosca *et al.* 1996). The lack of resuscitation with small sample size also indicates that only a small percentage of VBNC can actually be resuscitated.

4. Ingestion of VBNC *V. cholerae* by human volunteers resulted in the development of clinical symptoms of cholera, and the organism

could be isolated in culturable forms in stool. However, culturable cells could be isolated only on ingestion of cells that had been VBNC for 23 days, but not by those that had been VBNC for 4 weeks (Colwell *et al.* 1996; McDougald *et al.* 1998).

Only those findings under section A directly support the resuscitation of VBNC. Whereas those studies detailed under section B directly oppose the possibility of resuscitation of VBNC, all others indirectly suggest that resuscitation of VBNC may not have much significance, even though some authors still argue that they are significant.

Seasonality of cholera and the role of VBNC

The seasonal incidence of some infectious diseases has been observed for centuries. As a veterinary surgeon, I have observed the occurrence of bovine ephemeral fever, a viral disease affecting cattle, during the monsoon season in Kerala, the southernmost state in India. It appears soon after the onset of the monsoon (June) and remains for a few months, affecting a number of cattle, and then gradually disappears, only to reappear the following year. The same pattern was noticed during the four years (2002-2005) of my clinical experience at Parat, Kannur district, Kerala. Ephemeral fever is caused by a virus of *Rhabdoviridae* family, characterized by high fever, stiffness of limbs, lameness, anorexia, depression, drastic reduction in milk yield and nasal and ocular discharges (Walker 2005). Even though mortality is low, morbidity is very high, causing significant economic loss to farmers. What happens to the virus during the inter-epizootic periods? How does the virus reappear every monsoon? The virus is believed to be transmitted by biting insects like mosquitoes and midges, and the long-distance carriage of infected insects that appear to be borne on the wind is considered to be responsible for the spread of the disease (Yeruham *et al.* 2007; Walker 2005). Thus, insects act as vectors of the disease and, during the monsoon, when the population of these insects increases due to the favorable climatic and ecological conditions, the disease spreads rapidly. The incidence of the disease gradually reduces and later disappears. How and where the virus persists during the inter-epizootic periods is not clearly known. Since it may not remain in subclinical conditions in cattle, transovarial transmission through the vector is considered to be responsible for the persistence and reappearance of the disease every year (Yeruham *et al.* 2007). Thus, insects can act both as a vector and as a reservoir for ephemeral fever and a number of other infections. However, in the case of diseases like cholera, a live vector or reservoir may not be required.

Cholera is a devastating disease caused by the bacterium *Vibrio cholerae*, characterized by gastroenteritis and exhaustive diarrhea. The main pathogenic *Vibrios* in humans includes *V. cholerae*, *V. parahaemolyticus* and *V. vulnificus* (Daniels and Shafaie 2000; Hlady and Klontz 1996; Austin 2010). Transmission occurs mainly through the ingestion of food and water contaminated by the bacteria but also through the infection of open wounds (Austin 2010). Cholera is endemic to many Asian and African countries and predominant in South Asian countries like Bangladesh and India (Colwell 1996; Islam *et al.* 1993). Humans were originally thought to be the reservoirs of these bacteria, but research in the last few decades shows the aquatic environment as the main reservoir and that the bacterium can remain in water for long periods of time (Xu *et al.* 1982; Colwell 1996). An interesting feature of the organism is the difficulty of culturing it from aquatic samples during winter when the temperature goes below 10°C (Huq *et al.* 1990; Colwell 1996). However, after winter, when the temperature increases, they regain their culturability. How the bacteria that disappeared during the winter could reappear in summer was an enigma for a long time. With the discovery of Xu *et al.* (1982) that *Vibrio* can maintain a VBNC state at low temperatures and resuscitate on temperature upshift seemed to solve this enigma. Thus, it was proposed that the bacteria remain in culturable forms in aquatic environments during favorable conditions and assume a VBNC state during the winter, thus remaining viable year-round (Louis *et al.* 2003; Colwell 1996; Singleton *et al.* 1982; Binsztein *et al.* 2004; Gil *et al.* 2004; Lipp *et al.* 2002; Lipp *et al.* 2003). Huq *et al.* (1990) found that *V. cholerae* O1 could be detected throughout the year in the ponds and rivers of Bangladesh. Out of the 876 plankton samples collected, 63% were positive for the organism by fluorescent monoclonal antibody technique, whereas only 0.34% was detected by culture methods, indicating that most of the organisms remained in the VBNC state.

Temperature, salinity, and plankton population are the most important factors in the ecology of *V. cholerae* (Colwell 1996; Huq *et al.* 2005; Louis *et al.* 2003). Temperatures above 19°C and salinity between 2 and 14 ppt can be conducive for the multiplication of *V. cholerae* (Louis *et al.* 2003). Along with this, the phytoplankton blooms that occur during the warmer season followed by the increase in copepods provide a good source of nutrients for the organism (Colwell 1996). Many pathogenic *Vibrio* species are found to be attached to phytoplanktons, copepods (zooplanktons) and other marine organisms like crustaceans (Colwell 1996; Chowdhury *et al.* 1997; Rawlings *et al.* 2007; Grim *et al.* 2009; Thomas *et al.* 2006; Lipp *et al.* 2003). Many *Vibrio* species produce chitinase, attach to chitinaceous planktons and use chitin as the carbon source (Colwell 1996; Meibom *et al.* 2003). Water filtration using a simple method that employs a sari, a traditional dress of India and Bangladesh, can reduce the incidence of cholera drastically (Huq *et al.* 1996). The sari, a 5m-long piece of fabric, can be folded many times and can function as a good sieve, removing the planktons from the water and thus avoiding the bacteria attached to these organisms. Similarly, bathing and washing clothes in tube-well water was found to be more protective (Sack *et al.* 2003). This knowledge of the different climatic factors that are important in the etiology of cholera may be helpful in developing models of cholera outbreak using remote sensing techniques, integrating the environmental and biological data (Colwell 1996).

Frequent outbreaks of cholera that occur in Bangladesh and in some parts of India result from a number of precipitating factors including many climatic and socio-economic factors (Lipp *et al.* 2002; Gaffga *et al.* 2007; Greenstone 2009). High density of population, heavy rainfall, flood, pollution of water bodies, poverty, lack of toilet facilities, the unhygienic practice of passing stools in open land, lack of proper sewage, lack of clean drinking water etc. help spread the disease (Greenstone 2009; Colwell 1996; Gaffga *et al.* 2007; Sasaki *et al.* 2009; Akanda *et al.* 2009; Islam *et al.* 2009). The cholera incidences in Bangladesh show bimodal annual peaks, with the highest peak after the monsoon and a smaller peak during the spring (Lipp *et al.* 2002; Faruque *et al.* 2005; Akanda *et al.* 2009). These seasonal dual peaks can have separate reasons (Akanda *et al.* 2009). The first peak can be due to the heavy rain that washes all the sewage and waste into rivers and other water bodies, thus contaminating the source of drinking water (Akanda *et al.* 2009). During the spring, the low water levels in rivers may allow seawater to flow inland, thus bringing bacteria attached to the copepods and other planktons (Akanda *et al.* 2009). Thus both flood and drought can promote the transmission of cholera (Pascual *et al.* 2002; Koelle *et al.* 2005). The dual peaks in Bangladesh may also be related to the temperature and hours of daylight (Islam *et al.* 2009). High temperatures and medium daylight hours can provide favorable conditions for cholera transmission. During the summer, even though the temperature is high, the daylight hours can be shorter due to clouds, whereas during the winter, longer daylight hours maximize cholera incidence even though the temperature remains low (Islam *et al.* 2009). In some bi-annual seasonal cases of cholera, a dominant peak can be observed in the spring which can also be related to the surface temperature of the sea (Bouma and Pascal 2001). Thus a shift from the usual seasonal patterns is not uncommon, and the inter-annual variability may correspond to climate patterns in both long and short periods of time (Koelle *et al.* 2005). In Africa, South America, and most parts of the Indian subcontinent (other than the estuarine regions of Bangladesh, Bengal and Madras Province), cholera outbreaks usually show a single peak during the monsoon (Pascual *et al.* 2002).

By studying the seasonal patterns of cholera outbreak for 32 years in different parts of the world, Emch *et al.* (2008) found that the seasonal patterns are not apparent in regions near the equator, whereas they are pronounced at higher latitudes. Similarly, cholera outbreak is also more common in regions close to the equator, which can be attributed to the higher and constant temperatures (Emch *et al.* 2008; Pascual *et al.* 2002). Thus large climatic factors are important in the appearance and spread of the disease and can influence the strength and duration of seasonal patterns.

Seasonality of cholera is also influenced by *V. cholerae* bacterio-phages which may also help in the emergence of new pandemic serogroups or clones (Faruque *et al.* 2005). An inverse correlation was noticed between the phages and *V. cholerae* strains that are susceptible to phages in environmental water samples (Faruque *et al.* 2005). During the early periods of an epidemic, phage population is low, promoting bacterial multiplication. However, as the epidemic proceeds, phage population gradually increases, resulting in a lower bacterial population. Thus phages may play an important role in ending an epidemic (Faruque *et al.* 2005). During inter-epidemic periods, water samples mainly contain phages and no viable bacteria. However, during the next monsoon, heavy rain and flooding wash away the phages, providing a conducive environment for the remaining and freshly added bacteria to grow. Thus the phage predation can explain the seasonality of cholera epidemics (Faruque *et al.* 2005; Faruque *et al.* 2006a). Apart from the phages, bacteria are also ingested by protozoans (Matz *et al.* 2005; Arana *et al.* 2003).

The *V. cholerae* population is usually held in check by these predators (Worden *et al.* 2006). In a river microbial community, bacteria may be ingested by their predators even before they become VBNC cells (Arana *et al.* 2003). During phytoplankton blooms, *V. cholerae* may proliferate by overcoming predatory pressures (Worden *et al.* 2006).

Thus, *V.cholerae* is indigenous to riverine, estuarine and coastal waters and can be isolated either as free living bacteria or as those attached to planktons and other marine organisms. Occasional incidences of infections caused by *Vibrio* spp. have been reported from developed nations also (Normanno *et al.* 2005; Kirschner *et al.* 2008). In such cases, the main source of infection is the consumption of undercooked seafood or the infection of wounds after outdoor water contact (Normanno *et al.* 2005; Andersson and Ekdahl 2006). The occurrence of *Vibrio* spp. in the aquatic environment in the absence of human diseases is suggestive of the autochthonous nature of the organism (Colwell *et al.* 1981).

Global spread of *Vibrio*

Even though cholera and other diseases caused by *Vibrio* species are reported only rarely in developing countries, the organisms can be detected in aquatic environments worldwide (Emch *et al.* 2008). The global dissemination of *Vibrio* and the influence of climate are reviewed by Colwell (1996) and Nair *et al.* (2007). The seventh cholera pandemic that originated in Indonesia in 1961 spread to the Indian subcontinent and the Asian mainland during 1963-1969, to the Middle East and Africa in 1970-1971 and to South America in 1991 (Colwell 1996; Woodall 2009). In South America, it spread along the coastal regions for approximately 2000 km. The global spread is attributed to El Nino conditions, which may bring many climatic changes and associated phytoplankton blooms, thus facilitating the rapid growth and spread of the organism (Colwell 1996). Similarly, *V. parahaemolyticus* O3:K6, the causative agent of foodborne gastroenteritis, originated in Indonesia in 1995 (Nair *et al.* 2007). It then spread to India in 1996. The organism or its serovariants later spread to other parts of Asia, Africa, Europe and South and North America (Nair *et al.* 2007). In this case also, the rapid global spread of *V. parahaemolyticus* O3:K6 is attributed to El Nino conditions (Nair *et al.* 2007). Thus global temperature and other climatic factors can greatly influence the spread of *Vibrio* spp. Under certain conditions, *V. cholerae* remain in culturable form in seawater for sufficiently long periods of time to be carried by ocean currents to distant places (Colwell 1996). In spite of the ability of *Vibrio* to move to distant places, the seasonal cholera epidemic may not be due to the spread of a single clonal wave across a vast region (Stine *et al.* 2008). Instead, each site may have its own distinct group of strains, but one clone may have an advantage over others in causing the disease (Islam *et al.* 2004; Faruque *et al.* 2006b). This was evident from a study using different environmental and clinical strains of toxigenic *V. cholerae O1* isolated from four different geographic areas in Bangladesh (Islam *et al.* 2004).

The aquatic environment is considered to be the main reservoir for *Vibrio* spp (Colwell 1996). Apart from the marine, estuarine and riverine water, phytoplanktons, zooplanktons like copepods (Colwell 1996; Turner *et al.* 2009), sea weeds (Mahmud *et al.* 2006), algae (Hood and Winter 1997), cyanobacteria (Islam *et al.* 1999), aquatic plants like duckweed (Islam *et al.* 1990), crustaceans like oysters, mussels, shrimps and prawns (Ottaviani *et al.* 2009), fish (Senderovich *et al.* 2010), eels (Marco-Noales *et al.* 2004), sediments of lakes and ponds (Kaneko and Colwell 1973; Kaneko and Colwell 1975; Leal *et al.* 2008), protozoan *Acanthamoeba castellanii* (Abd *et al.* 2007) and biofilms in aquatic systems (Alam *et al.* 2007) can act as reservoirs for various *Vibrio* species. Aeroplanktons (Paz and Broza 2007), chironomid eggs (Halpern *et al.* 2004; Halpern *et al.* 2007), abiotic surfaces (Snoussy *et al.* 2008), cargo ship ballast (McCarthy and Khambaty 1994), fish (Senderovich *et al.* 2010), migratory water birds (Halpern *et al.* 2008), and movement of people (Kiehl 1998; Strumbelj *et al.* 2005; Strauss *et al.* 2004; O'Brien and Stanwell-Smith 1998) can help in the long-distance spread of cholera, even between the continents. Similarly, food supplies shipped internationally can carry the organism from the original endemic location (Taylor *et al.* 1993). In 1991, three cases of cholera infection were reported in Maryland, U.S.A., after the consumption of frozen coconut milk, produced and imported from Thailand (Taylor *et al.* 1993). People with asymptomatic subclinical infections may shed the organism in the stool and thus can be yet another important carrier in the spread of cholera (King *et al.* 2008).

Hosts can also induce the epidemic spread of cholera since the colonization of bacteria in the gastrointestinal tract of humans creates a hyperinfectious state, resulting in a competitive advantage for those bacteria in the stool (Merrell *et al.* 2002). The hyperinfectious bacteria exhibit high motility and increased ability to acquire nutrients, but have lower bacterial chemotaxis, as evident from the expression levels of those genes regulating motility and chemotaxis (Merrell *et al.* 2002). The increased infectivity of the newly shed bacteria can be due to the ability of the bacteria to store glycogen, which can prolong their survival in the aquatic environment that is poor in nutrients (Bourassa and Camilli 2009). The presence of glycogen granules in the stools of cholera patients also indicate that glycogen may contribute to the pathogenesis of *V. cholerae* by facilitating its transmission (Bourassa and Camilli 2009). Faruque *et al.* (2006a) suggested that the high infectivity of *V. cholerae* shed in human stools can be due to the ability of the bacteria to develop into *in vivo*-formed biofilms. Freshly shed stools contain both planktonic bacteria and biofilm-like structures with cell aggregates, which may provide a sufficiently high dose of the pathogen to cause infection (Faruque *et al.* 2006a). Once the newly shed bacteria enter the aquatic environment, they may not maintain their hyperinfectious state for long (Faruque *et al.* 2006a). The ability of bacteria to colonize the human intestine may not be affected immediately after shedding from the patients, but may reduce after 24 h of incubation in pond water (Nelson *et al.* 2008). Hence a rapid transfer to the next host may be required to maintain a fitness advantage for *V. cholerae* to overcome the negative selection pressures in the aquatic environment (Nelson *et al.* 2008).

Conclusion

Bacteria enter into the VBNC state during unfavorable conditions where they remain viable but are non-culturable and exhibit very low levels of metabolic activity. The ability of bacteria to enter the VBNC state is considered to be a survival strategy of the bacterial population against starvation and stress. When the conditions become favorable, VBNC cells resuscitate and return to culturability. The absence of *V. cholerae* in water samples during winter and their reappearance during summer is attributed to their ability to enter the VBNC state at low temperatures. Thus, resuscitation of VBNC can be responsible for the seasonal outbreak of cholera in endemic areas. Similarly, *V.cholerae* is indigenous to riverine, estuarine and coastal waters and is present in aquatic environment all through the year, either as culturable or as VBNC cells.

Section 2
VBNC: A Tale of Two Illusions

U nlike persisters and SCVs, the nomenclature and the significance of VBNC have been criticized by many researchers. In this section, I discuss why VBNC is not a successful phenotype and is not responsible for either seasonal outbreak or the global spread of cholera and other diseases.

1. Resuscitation of VBNC is not predictable.

A number of researchers have proposed that VBNCs can be resuscitated *in vitro, in vivo* and *in situ*. However, many other researchers have failed to resuscitate VBNC cells (as reviewed in the previous section). Thus it is impossible to predict whether a given VBNC retains the ability to resuscitate or not. There is no 'VBNC marker' that can predict the resuscitability of VBNC. Thus, the most important factor in an argument against the significance of VBNC is the lack of predictability of its ability to resuscitate.

2. Storage of bacterial culture at 4°C for prolonged periods is not recommended.

For laboratory works, many bacterial cultures are stored at 4°C for short-term storage of a few days or weeks and at -80°C for long-term storage. At 4°C, bacteria may remain viable for 1-2 months but may lose their culturability on long-term storage. The loss of culturability is true for both stationary and exponential phase bacteria, even in the nutrient-rich LB medium. At -80°C, they maintain their culturability for many months or even years and do not require any special resuscitation techniques. If bacterial cultures stored at 4°C for long periods can remain in the VBNC state and can later be resuscitated, why do we need to keep bacterial cultures at -80°C for prolonged storage? The next statement may have the answer.

3. VBNC cells do not resuscitate after prolonged periods of exposure to stressors.

The ability of VBNCs to resuscitate depends on the total time of exposure to stressors. For example, copper-induced *Escherichia coli* VBNC cells could be resuscitated for only up to 2 weeks after entering the VBNC state (Grey and Steck 2001b). Thus, if bacteria have entered the VBNC state recently, it may be possible to resuscitate some of them. However, if the exposure time is prolonged, resuscitation may not be possible by any methods, even if they appear viable.

4. Not all VBNC cells maintain their ability to resuscitate.
Among the VBNC cells, only a small population, possibly of the order one in 10^3 or 10^4 cells, may resuscitate while the majority of the population cannot be resuscitated even though they appear viable (Coutard *et al.* 2007; del Campo *et al.* 2009; Lleo *et al.* 2001; Weichart *et al.* 1997). One can argue that this is an altruistic behavior and a survival strategy exhibited by the bacterial population, wherein only a small subpopulation is maintained under nutrient limitation at the expense of the majority. However, VBNCs themselves can be considered a survival strategy, as they adopt a lower metabolically active state that does not require many nutrients. Hence most of the VBNC cells should be capable of resuscitation upon addition of nutrients if they were genetically programmed. The ability of a small subpopulation of VBNC cells to resuscitate can be due to cryptic growth (maintenance of life from the nutrients released by dead bacteria) which can be a random process rather than an altruistic behavior.

5. Recovery of growth after temperature upshift can be due to the regrowth of culturable bacteria.
In 1982, Colwell's lab reported that *V. cholerae* and *E. coli* can remain in the VBNC state (Xu *et al.* 1982), which later led to the proposal that VBNCs can resuscitate when conditions become favorable. However, after a decade, the same lab reported that, following temperature upshift, true resuscitation from VBNC does not occur in either *E. coli* or *V. cholerae* and that the growth recovery could be due to the regrowth of a few culturable cells only (Chowdhury *et al.* 1994; Ravel *et al.* 1995), thus raising questions about the significance of VBNC.

6. Starvation or other stressors may not lead to an abrupt end to bacterial life.
Starvation or temperature stress can lead to gradual cellular deterioration. Initially some cells die, followed by a gradual reduction in the number of culturable bacteria and later leading to the loss of culturability of most of the population. During these processes, respiratory activity and enzymatic activity gradually reduces and finally disappears. Degradation of RNA, DNA and cell membrane accumulates (Yamamoto 2000). When the stress is prolonged, cellular deteriora-

tion continues and finally death occurs. Some of the vital responses of cell may not be the systematic signs of life (Yamamoto 2000) and hence the cells may appear viable when certain techniques for determining cell viability are used, even though they had undergone an irreversible loss of culturability. Some sublethally injured cells (depending on the stage of deterioration) may resuscitate upon the removal of stress and the addition of nutrients before complete loss of culturability. The fact that only a small percentage of VBNC cells can be resuscitated and that such cells cannot be resuscitated after prolonged exposure to stressors indicate that the VBNC state may not be a genetically programmed state but rather a 'gradual cellular death response.'

7. The incidence of cholera is low during inter-epidemic periods.
During the winter, when the incidence of cholera is low, *Vibrio* can be detected in aquatic environments mainly in the VBNC state. However, the incidence of cholera should still remain at high levels if VBNC cells maintain their ability to resuscitate inside the human body. Researchers had earlier proposed that a simple temperature upshift alone is sufficient for the resuscitation of *V. vulnificus* VBNC (Oliver *et al.* 1995). Similarly, a number of experiments have proved the ability of VBNCs to resuscitate *in vivo* (reviewed earlier). If the proponents of VBNC resuscitation are correct, most VBNCs should be able to resuscitate and cause infection in humans following exposure to higher body temperatures.

8. *Vibrio* species can be isolated in culturable forms even during winter.
Even though the isolation of culturable *Vibrio* spp. during the winter is difficult, there are number of reports wherein they were isolated both from the natural environment and under experimental conditions at low temperatures.

a. High counts of *V. cholerae* were noticed from the polluted water samples in Calcutta even during winter (Nair *et al.* 1988).

b. Culturable *V. cholerae* 0139 could be isolated from aquatic environments (water, fish, snail and duckweed) in Bangladesh during winter (Islam *et al.* 1996).

c. Even though the number of culturable *V. cholerae* was greater during warmer months, they could also be isolated during the winter from the water and plankton samples of the Chesapeake Bay (Louis *et al.* 2003).

d. *V. cholerae* was isolated from the cloacal swabs of gulls during the time when the organism could not be detected in water (Lee *et al.* 1982).

e. Even though the water temperature and the number of *V. parahaemolyticus* isolated from Alabama oyster samples were highly correlated, the organism could be isolated even during winter when the temperature was 10°C. In fact, pathogenic strains were more prevalent in the winter (DePaola *et al.* 2003). Similarly, even though the densities of *V. vulnificus* were low during the winter, the organism could be isolated from the intestines of many estuarine fishes (DePaola *et al.* 1994).

f. Oysters artificially contaminated with *V. vulnificus* survived storage at 10°C and below. Similarly, in uninoculated control oysters, endogenous *V. vulnificus* were found after 7 days at both 0.5 and 10°C. In addition, oysters that had taken up *V. vulnificus* from seawater retained the bacterium for 14 days at 2°C (Kaysner *et al.* 1989).

g. *V. vulnificus* survived in oysters at low temperatures for many days even though their numbers reduced with decreasing temperatures. At temperatures between 0.5-6°C, the number reduced to 0-8% of the initial concentration in 13 days. However, at 8.5-10.5°C, their number remained high for the whole duration of experiment (Kaspar and Tamplin 1993). Thus, even at low temperatures, the culturable organism can be isolated from oysters.

h. *V. vulnificus* could be detected in seawater and oysters at water temperatures of 8.0 and 7.68°C respectively (Wright *et al.* 1996).

i. Many *Vibrio* species could be isolated from the hemolymph of blue crabs from Galveston Bay, Texas, collected during different seasons (Davis and Sizemore 1982). Similarly, hemolymph and soft tissues of Pacific oysters kept in seawater at 1-8°C were found to contain many bacteria, including *Vibrio* (Olafsen *et al.* 1993).

j. *V. vulnificus* could be isolated from the coastal waters in South Texas throughout the year at a concentration of 3.9 × 10³ cfu/100 ml, even at water temperature of 9·7°C (Ramirez *et al.* 2009).

k. *V. vulnificus* were isolated from the water samples in a mussel farm in Denmark at a temperature of 7°C (Hoi *et al.* 1998).

l. Relatively high numbers of viable *V. cholerae* could be isolated from the ice samples used for making fruit juice obtained from street vendor stalls in Mumbai, India (Mahale *et al.* 2008).

m. Culturable *Vibrio* species could be isolated from water and plankton samples throughout the year (even during winter when the temperature was 9.8°C) from the coastal waters of Georgia, USA (Turner *et al.* 2009).

The above findings indicate that culturable bacteria are still present in water, planktons or other marine organisms (albeit at low concentrations) during the winter, when the temperature is very low. But as the temperature increases and the plankton boom occurs during the summer, the few culturable bacteria may begin to multiply, increasing the total number of bacteria. Thus, *Vibrio* concentrations may be directly related to the temperature and other environmental factors, but the increase could be due to the few culturable bacteria itself rather than due to VBNC.

9. *V. shiloi* VBNCs may not have any significance in coral bleaching.

V. shiloi is considered to be the causative agent of the bleaching of coral *Oculina patagonica* (Kushmaro *et al.* 2001). During the summer, *V. shiloi* may penetrate and infect the corals, multiplying intracellularly, where most of the bacteria remain in the VBNC state (Israely *et al.* 2001). During the winter, when the temperature of sea water falls below 20°C, they may remain viable outside the corals, but not inside the healthy or bleached corals. The absence of *V. shiloi* in corals during the winter and their intracellular killing suggests that during each spring, a fresh infection rather than the reactivation of VBNCs in the corals is responsible for the summer bleaching (Israely *et al.* 2001). Later, it was found that the marine fireworm *Hermodice caranculata*, acts as a winter reservoir for *V. shiloi*, where more than 99.9% of the bacteria remain in the VBNC state (Sussman *et al.* 2003). Worms taken from the sea during the winter contained approximately 10^8 *V. shiloi* per worm, of which approximately 10^4 could form colonies. When the worms were infected with *V. shiloi*, the bacteria adhered and penetrated into the worm; those worms could infect the corals and cause bleaching, indicating that *H. caranculata* acts as a vector for transmitting the bleaching disease (Sussman *et al.* 2003; Rosenberg and Falkovitz 2004).

The above findings indicate that VBNCs inside *O. patagonica* were not responsible for reinfections during the summer as they were killed at low temperatures. Given the fact that approximately 10^4 bacteria in the worm could form colonies, it can be assumed that those bacteria were responsible for reinfection rather than VBNCs (even though a vast majority of bacteria remain in the VBNC state). It is still not known whether bacteria constitute the agent responsible for coral bleaching (Ainsworth *et al.* 2008), but if it is indeed due to bacteria, it can be due to culturable bacteria itself rather than VBNCs.

10. *Vibrio* spp. are regularly introduced into the aquatic environment.

Even though *Vibrio* species are autochthonous to aquatic environments, new strains or prevailing ones are continuously introduced into the environment in different ways. Paz and Broza (2007) reported that cholera could spread with the prevailing winds and that adult non-biting midges could carry *V. cholerae* from one body of water to another. Chironomid (non-biting midges that are widely distributed in freshwater) egg masses can act as a reservoir for *V. cholerae* (Halpern *et al.* 2007; Broza *et al.* 2008: Broza *et al.* 2005). Most of the *V. cholerae* in the egg masses were in the VBNC state, but a number of culturable bacteria could also be isolated (Halpern *et al.* 2007). Chironomid populations peak biannually, and it was noticed that the peaks were followed by subsequent bacterial growth during the summer and disappearance during the winter (Halpern *et al.* 2004). Chironomids can act as both a reservoir and a carrier of *V. cholerae*. However, long-distance dispersal of *Vibrio* may not be possible based on the movement of chironomids with the wind. But it is possible that migratory birds like waterfowl that feed on copepods and chironomid eggs can disseminate the organism within and between the continents (Halpern *et al.* 2008). Fecal samples of many aquatic birds in Colorado and Utah were found to contain both O1 and non-O1 *V. cholerae* strains (Ogg *et al.* 1989). Similarly, *V. cholerae* could also be isolated from the gut of many freshwater fish (Halpern *et al.* 2008). Fish that consume copepods and chironomids can act as the intermediate reservoir for *Vibrio* (Senderovich *et al.* 2010). Some marine fishes can migrate to distant places following temperature variations in the ocean and thus have the potential to carry the organism to different places (Senderovich *et al.* 2010).

Cargo ship ballast water taken from cholera-infected countries and released into the coastal waters at various ports can also be a source of transmission of *Vibrio* (McCarthy and Khambaty 1994). In fact, ship ballast water is considered to be the source of transmission of the seventh cholera pandemic to South America (McCarthy and Khambaty 1994). Similarly, the movement of people (Kiehl 1998; Strumbelj *et al.* 2005; Strauss *et al.* 2004; O'Brien and Stanwell-Smith 1998) and the import of seafood between different continents (Ottaviani *et al.* 2009) can also help in the transmission of disease. Warm-blooded animals can be yet another reservoir for the organism (Visser *et al.* 1999; DePaola *et al.* 2003). Raccoons that feed on oysters may harbor high densities of *V. parahemolyticus* in their intestines even during winter (DePaola *et al.* 2003). Similarly, human-associated activity can also be a very important source of transmission of *Vibrio*, which is given below.

Thus, the outbreak of cholera in a particular area may not always be due to the regrowth of autochthonous (indigenous or native) *V. cholerae*, but can be due to the newly introduced organisms from neighboring or distant places.

11. Human-associated sources of infection may be more important than environmental sources of infections.

Sewage and fecal contamination of drinking water and the ingestion of contaminated food are considered the most important source of transmission for cholera. Passing stools in open places, insufficient drainage systems etc. result in the contamination of drinking water sources which, following heavy rain, add bacteria to aquatic environments. Persons with acute cholera may excrete 10^7-10^8 *V. cholerae* per gram of stool (Kaper *et al.* 1995). Untreated patients and convalescent patients may excrete *Vibrio* for 1-2 weeks, whereas a small minority of the population may continue excreting them for much longer periods (Kaper *et al.* 1995). Similarly, patients with immuno-deficiency diseases may excrete *Vibrio* for longer periods (Utsalo *et al.* 2004), a factor critical in high-risk areas for HIV and cholera (von Seidlein *et al.* 2008). Moreover, a high percentage of people in endemic areas may have inapparent infections (King *et al.* 2008; Puglielli *et al.* 1992; Kaper *et al.* 1995; Olivier *et al.* 2007). It had been believed that human asymptomatic and convalescent patients were the major reservoirs of cholera. However, with the finding that the bacteria can remain in water as VBNCs for a long time, the aquatic environment has been widely considered to be the major reservoir of *Vibrio* species. But, an article published in *Nature* recently (King *et al.* 2008) indicates that human-associated cholera infection can be more important than the environmental source of infection. Similarly, the asymptomatic ratio can be very high in cholera, and environmental reservoirs may be directly responsible for only a few infections (King *et al.* 2008). This can be especially true for *V. cholerae* O1 El Tor which produces more asymptomatic infections than the *V. cholerae* O1 classical strain and exhibits prolonged periods of *Vibrio* excretion (Olivier *et al.* 2007; Woodward *et al.* 1972). These factors might have contributed to the dissemination of the El Tor strain worldwide by facilitating person-to-person transmission. Thus, the disease outbreak pattern may depend on the prevalence of inapparent infections (King *et al.* 2008). Isolation of *V. cholerae* O1 from 3% of the total randomly selected, symptom-free individuals with no signs of cholera or diarrhea during the seventh pandemic cholera infection in Peru in 1991 (Puglielli *et al.* 1992) also indicates the importance of inapparent infections in the spread of disease.

12. Spread of seventh pandemic *V. cholerae* may not have involved VBNC (or even culturable autochthonous *V. cholerae*). The seventh pandemic of cholera had four phases (Kaper *et al.* 1995; Woodall 2009; Colwell 1996). Unlike the previous six pandemics that originated in South Asia, the seventh pandemic originated in Sulawesi, Indonesia in 1961. During the first phase (1961-1962), it spread to other islands in Indonesia, the Philippines and Taiwan, thus affecting many regions in Southeast Asia. In the second phase (1963-1969), it affected Malaysia and then spread to Cambodia, Vietnam, Burma and Bangladesh. This was followed by the organism spreading to India, Pakistan, Afghanistan, Iran, Iraq and some parts of Soviet Union. In the third phase (1970-1971), it reached most of the Middle East and West Africa. From West Africa, it spread to sub-Saharan areas along the coast and into the interior along the rivers. Out of the 36 countries affected by cholera in 1970, 28 were newly affected countries. In the fourth phase (1991-1992), it reached Peru, South America, 20 years after Africa was affected. It started in three different areas in Peru simultaneously, separated by hundreds of kilometers. This was the first time cholera had entered South America for nearly a century. From Peru, it rapidly spread to most of the countries in South America, affecting a total of about 750,000 people and killing 6,500 (Kaper *et al.* 1995). However, within a decade, cholera was almost eliminated from South America (Woodall 2009).

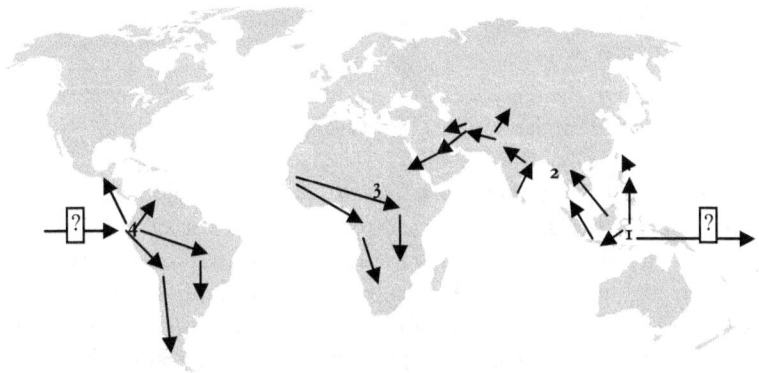

Fig. 2. Global spread of seventh pandemic of cholera

How did the seventh pandemic strain reach Peru? The origin of this strain in Peru is not clear even now. However, there is much speculation (Seas *et al.* 2000; Wachsmuth *et al.* 1993; McCarthy and Khambaty 1994), including four main theories.

1. Non-toxigenic *V. cholerae* O1 already present in the aquatic environment of Peru acquired virulence through recombination events (Karaolis *et al.* 1995). This would suggest that the seventh pandemic in Peru evolved independently and is unrelated to the strain from the rest of the world.

2. Toxigenic *V. cholerae* O1 was already present in the coastal waters of Peru either as plankton-associated VBNC or as culturable bacteria at low concentrations. Isolation of toxigenic *V. cholerae* from the coastal waters of Peru during the pandemic period made some researchers speculate that *V. cholerae* autochthonous to aquatic environments of Peru was the source of origin of cholera there (Lipp *et al.* 2003; Seas *et al.* 2000; Gil *et al.* 2004). As per Lipp *et al.* (2003), *"It is highly unlikely that cholera in Peru arrived from other cholera-affected regions of the world. A more plausible explanation is that V. cholerae is autochthonous to Peruvian coastal, brackish, and riverine waters...."*

3. The new strain was carried from endemic areas either by travelers or by cargo ships that released ballast water into the ports of Peru, thus contaminating the seawater and seafoods (Wachsmuth *et al.* 1993; McCarthy and Khambaty 1994). Lack of evidence of *V. cholerae* O1 infection in Peru and Mexico City before the outbreak of pan-

demic infection is suggestive of introduction from other regions (Wachsmuth *et al.* 1993).

4. Currents of the Pacific and Indian Oceans carried toxigenic *V. cholerae* O1 to Peruvian coasts (Mourino-Perez 1998).

The sixth pandemic was caused by *V. cholerae* O1 classical biotype whereas the seventh was caused by *V. cholerae* O1 El Tor biotype. Latin American isolates are considered to be the variants of the seventh pandemic clone (Wachsmuth *et al.* 1993). By comparing *V. cholerae* O1 El Tor strains of Latin America of the seventh pandemic with the strains of the seventh pandemic from the rest of the world, Latin American strains were found to be the variants of Asian/African strains that had evolved independently (Wachsmuth *et al.* 1993; Nusrin *et al.* 2009). Chun *et al.* (2009) proposed that the sixth pandemic strain had undergone a 'shift' to give rise to the seventh pandemic strain, whereas the seventh Asian pandemic strain had undergone a 'drift' to give rise to diverse variants. As per this proposal, a non-toxigenic precursor for the Latin American strain is unlikely. However, this is in contrast with the findings of Karaolis *et al.* (1995), who proposed that the sixth and the seventh pandemics were independent clones derived from the nontoxigenic, environmental, non-O1 *V. cholerae* isolates. Thus, as per Karaolis *et al.* (1995), it is possible to have a non-toxigenic precursor for the Latin American strains. Variation in the Latin American strains also led to the proposal that ocean currents were responsible for carrying the strain across the Pacific (Mourino-Perez 1998). It was argued that the time *V. cholerae* had taken to reach Latin America (30 years) and the different environment it had encountered during the journey could explain why the Latin American strains were different from the pandemic strain isolated from the rest of the world (Mourino-Perez 1998).

An interesting feature of the *V. cholerae* O1 El Tor outbreak in Africa is that it started in 1970 in West Africa at Guinea-Bissau, probably from a returning traveler (Kaper *et al.* 1995), rather than in East Africa as a continuation from the Middle East. Similarly, a unique ribotype of *V. cholerae* O1 El Tor that appeared in Calcutta, India, in September1993, also reached the West African coast at Guinea-Bissau in October 1994 and lasted into 1996 (Sharma *et al.* 1998). Even though the source of cholera outbreak in Africa in both cases is not known, it is possible that both the strains reached West Africa from India.

One of the arguments against the proposal that autochthonous *V. cholerae* is the source of the cholera outbreak in Peru is the failure to detect cholera in Latin America before the outbreak (Varela *et al.* 1971; Wachsmuth *et al.* 1993). Even though toxigenic *V. cholerae* could be isolated from seawater during the pandemic period, cholera infection was not present on the continent for nearly a century (Colwell 1996), and no toxigenic strains were detected before the outbreak (Wachsmuth *et al.* 1993). Another factor is the introduction of the new strain into 28 newly affected countries out of the 36 countries affected by the seventh pandemic strain (Kaper *et al.* 1995). It is easily noticeable that the spread of the seventh pandemic of cholera had a definite direction; that is, it started from Indonesia, spread along the coast and then entered the mainland of most Southeast Asian countries, followed by the Indian subcontinent, the Middle East and then Africa in a period of 10 years. It is clear that the outbreak of cholera in these regions was indeed due to a 'gradual spread' rather than an 'emergence' from any indigenous strains. If the seventh cholera pandemic had emerged from autochthonous *V. cholerae* already present in the aquatic environment, one would expect outbreaks of cholera from multiple regions on different continents. The speculation, therefore, that the autochthonous *V. cholerae* was responsible for the cholera outbreak in Peru seems to be less likely. Similarly, the fact that the outbreak of cholera occurred at Guinea-Bissau in West Africa in both 1970 and 1994 and that both the strains were earlier detected in Calcutta, India, indicates that they could have been introduced to West Africa from Calcutta (Sharma *et al.* 1998), possibly through a common unknown carrier. Even though the origin of cholera in Peru is still unknown, it could also have reached the area with the help of carriers from other continents rather than through the ocean currents in the Pacific.

The significance of imported *V. cholerae* is highlighted by the incidence of cholera in three women in Sydney, Australia (Frossman *et al.* 2007). The patients had no history of travel to cholera-endemic areas, and an investigation showed that they got infected after consumption of raw whitebait imported from Indonesia (Frossman *et al.* 2007). Thus it is possible that imported seafoods can be a source of outbreak in cholera-free regions and help in its global spread (Minami *et al.* 1991; Chen *et al.* 2004; Frossman *et al.* 2007).

Global spread of *V. parahaemolyticus* O3:K6

Gastroenteritis due to *V. parahaemolyticus* occurs mainly due to consumption of contaminated raw or improperly cooked marine foods (Barker and Gangarosa 1974; Su and Liu 2007). The organism is widely disseminated throughout the world in marine and estuarine environments (Nair *et al.* 2007). Many gastroenteritis cases were reported in Calcutta, India, in February 1996, caused by a unique serotype *V. parahaemolyticus* O3:K6 (Nair *et al.* 2007). However, the first O3:K6 isolate was isolated from a traveler in Japan in 1995 who was returning from Indonesia (Nair *et al.* 2007). It is interesting to note that the seventh pandemic *V. cholerae* also originated in Indonesia (Kaper *et al.* 1995; Colwell 1996). From 1996 onwards, a number of gastroenteritis cases due to the O3:K6 isolate were reported in many South East Asian countries including Japan, Taiwan, Thailand, and Korea, as well as in Bangladesh (Nair *et al.* 2007). It was reported in the United States in 1998 and was isolated from persons from different states who had eaten oysters harvested from Galveston Bay, Texas (Nair *et al.* 2007). This serotype was not detected in the U.S. before the outbreak. Even though *V. parahaemolyticus* infections were reported in Spain before 1996, it was mainly due to the serotype O4:K11 (Martinez-Urtaza *et al.* 2004). However, in 2004, infection due to the O3:K6 serotype was reported in Spain and France (Martinez-Urtaza *et al.* 2004; Quilici *et al.* 2005). Outbreaks caused by the O3:K6 serotype occurred in Chile and Peru between 1997 and 1998 (Nair *et al.* 2007). They were also reported from Russia, China and Mozambique (Nair *et al.* 2007). Thus, the O3:K6 isolate had spread to at least 5 different continents. How this isolate spread to different parts of the world is not known. It might have reached either through cargo ship ballast water (McCarthy and Khambaty 1994) or through imported seafoods (Matsumoto *et al.* 2000). El Nino is also considered responsible for the spread of disease (Colwell 1996; Nair *et al.* 2007). Since this isolate was not reported in the U.S. before the outbreak, the possibility that the organism was imported from other endemic areas is very high. This can be true for the seventh pandemic *V. cholerae* O1 El Tor outbreak in Peru also, even though some researchers suggest that it is unlikely that cholera in Peru arrived from other cholera-affected regions of the world (Lipp *et al.* 2003).

Conclusion

Until now, there has been no convincing evidence that VBNCs are responsible for seasonal cholera outbreaks. It is also not known whether the growth recovery of a bacterial culture exposed to various stressors is due to the resuscitation of VBNC or the re-growth of a few culturable bacteria. It is interesting to note that the same group that proposed the resuscitation of VBNCs has offered many explanations for growth recovery following temperature upshift:

1. Resuscitation of VBNC in a favorable climate is responsible for seasonal outbreak of cholera (Colwell 1996)

2. Re-growth of a few culturable bacteria, and not the resuscitation of VBNC, is responsible for the growth recovery of *V. cholerae* and *E. coli* (Chowdhury *et al.* 1994; Ravel *et al.* 1995).

3. After temperature upshift, the re-growth of a portion of VBNC cells of *V. parahaemolyticus* subjected to cold temperatures is responsible for the recovery of culturability of VBNC cells, rather than the resuscitation of all bacteria of the initial inoculum (Coutard *et al.* 2007).

Thus, it is still not clear whether the resuscitation of VBNC is true and has any significance.

Similarly, the pattern of global spread of *V. cholerae* O1 or *V. parahaemolyticus* O3:K6 suggest that VBNC or even autochthonous *Vibrio* organisms may not have much role in the spread of cholera outbreaks worldwide. Rather than originating from multiple regions in different continents, it might have originated in a single/few area(s) and then spread to different parts of the world. There is no denying that the aquatic environment plays an important role in the ecology of *V. cholerae* in endemic areas. Water, phytoplanktons, zooplanktons, etc. are important reservoirs for the organism. However, this alone may not be sufficient to explain the spread of *Vibrio* to different continents. Other carriers such as ship ballast water, movement of people, migratory birds, chironomids, imported seafoods and asymptomatic human carriers may also be important in the ecology of *V. cholerae*. Similarly, the endemicity of cholera may depend more on the human activity than on environmental factors.

In short, Colwell's group has put forward three theories:

i. resuscitation of VBNC is responsible for the seasonal outbreak of cholera in endemic areas;

ii. the aquatic environment plays an important role in the ecology of *V. cholerae*, and

iii. autochthonous *V. cholerae* is responsible for the presence of pandemic strains across the world.

However, out of these theories, both the first and the third theories remain scientifically questionable.

References

Abd, H., Saeed, A., Weintraub, A., Nair, G. B., and Sandstrom, G. (2007). Vibrio cholerae O1 strains are facultative intracellular bacteria, able to survive and multiply symbiotically inside the aquatic free-living amoeba Acanthamoeba castellanii. *Fems Microbiology Ecology* 60(1), 33-39.

Ainsworth, T. D., Fine, M., Roff, G., and Hoegh-Guldberg, O. (2008). Bacteria are not the primary cause of bleaching in the Mediterranean coral Oculina patagonica. *ISME J* 2(1), 67-73.

Akanda, A. S., Jutla, A. S., and Islam, S. (2009). Dual peak cholera transmission in Bengal Delta: A hydroclimatological explanation. *Geophysical Research Letters* 36, -.

Alam, M., Hasan, N. A., Sadique, A., Bhuiyan, N. A., Ahmed, K. U., Nusrin, S., Nair, G. B., Siddique, A. K., Sack, R. B., Sack, D. A., Huq, A., and Colwell, R. R. (2006a). Seasonal cholera caused by Vibrio cholerae serogroups O1 and O139 in the coastal aquatic environment of Bangladesh. *Applied and Environmental Microbiology* 72(6), 4096-4104.

Alam, M., Sultana, M., Nair, G. B., Sack, R. B., Sack, D. A., Siddique, A. K., Ali, A., Huq, A., and Colwell, R. R. (2006b). Toxigenic Vibrio cholerae in the aquatic environment of Mathbaria, Bangladesh. *Applied and Environmental Microbiology* 72(4), 2849-2855.

Alam, M., Sultana, M., Nair, G. B., Siddique, A. K., Hasan, N. A., Sack, R. B., Sack, D. A., Ahmed, K. U., Sadique, A., Watanabe, H., Grim, C. J., Huq, A., and Colwell, R. R. (2007). Viable but nonculturable Vibrio cholerae o1 in biofilms in the aquatic environment and their role in cholera transmission. *Proceedings of the National Academy of Sciences of the United States of America* 104(45), 17801-17806.

Allegra, S., Berger, F., Berthelot, P., Grattard, F., Pozzetto, B., and Riffard, S. (2008). Use of Flow Cytometry To Monitor Legionella Viability. *Applied and Environmental Microbiology* 74(24), 7813-7816.

Alleron, L., Merlet, N., Lacombe, C., and Frere, J. (2008). Long-Term Survival of Legionella pneumophila in the Viable But Nonculturable State After Monochloramine Treatment. *Current Microbiology* 57(5), 497-502.

Andersson, Y. a. E., K. (2006). Wound infections due to *Vibrio cholerae* in Sweden after swimming in the Baltic sea, summer 2006. *Euro Surveill* 11(31), 3013.

Arana, I., Irizar, A., Seco, C., Muela, A., Fernandez-Astorga, A., and Barcina, I. (2003). gfp-tagged cells as a useful tool to study the survival of Escherichia coli in the presence of the river microbial community. *Microbial Ecology* 45(1), 29-38.

Arana, I., Orruno, M., Perez-Pascual, D., Seco, C., Muela, A., and Barcina, I. (2007). Inability of Escherichia coli to resuscitate from the viable but nonculturable state. *Fems Microbiology Ecology* 62(1), 1-11.

Austin, B. (2010). Vibrios as causal agents of zoonoses. *Veterinary Microbiology* 140(3-4), 310-317.

Baffone, W., Casaroli, A., Citterio, B., Pierfelici, L., Campana, R., Vittoria, E., Guaglianone, E., and Donelli, G. (2006). Campylobacter jejuni loss of culturability in aqueous microcosms and ability to resuscitate in a mouse model. *International Journal of Food Microbiology* 107(1), 83-91.

Baffone, W., Citterio, B., Vittoria, E., Casaroli, A., Campana, R., Falzano, L., and Donelli, G. (2003). Retention of virulence in viable but non-culturable Vibrio spp. *International Journal of Food Microbiology* 89(1), 31-39.

Barer, M. R. (1997). Viable but non-culturable and dormant bacteria: Time to resolve an oxymoron and a misnomer? *Journal of Medical Microbiology* 46(8), 629-631.

Barker, W. H., and Gangaros.Ej (1974). Food Poisoning Due to Vibrio-Parahaemolyticus. *Annual Review of Medicine* 25, 75-81.

Basaglia, M., Povolo, S., and Casella, S. (2007). Resuscitation of viable but not culturable Sinorhizobium meliloti 41 pRP4-luc: Effects of oxygen and host plant. *Current Microbiology* 54(3), 167-174.

Beumer, R. R., Devries, J., and Rombouts, F. M. (1992). Campylobacter-Jejuni Nonculturable Coccoid Cells. *International Journal of Food Microbiology* 15(1-2), 153-163.

Binsztein, N., Costagliola, M. C., Pichel, M., Jurquiza, V., Ramirez, F. C., Akselman, R., Vacchino, M., Huq, A., and Colwell, R. (2004). Viable but nonculturable Vibrio cholerae O1 in the aquatic environment of Argentina. *Applied and Environmental Microbiology* 70(12), 7481-7486.

Biosca, E. G., Amaro, C., MarcoNoales, E., and Oliver, J. D. (1996). Effect of low temperature on starvation-survival of the eel pathogen Vibrio vulnificus biotype 2. *Applied and Environmental Microbiology* 62(2), 450-455.

Bogosian, G., Aardema, N. D., Bourneuf, E. V., Morris, P. J. L., and O'Neil, J. P. (2000). Recovery of hydrogen peroxide-sensitive culturable cells of Vibrio vulnificus gives the appearance of resuscitation from a viable but nonculturable state. *Journal of Bacteriology* 182(18), 5070-5075.

Bogosian, G., and Bourneuf, E. V. (2001). A matter of bacterial life and death. *Embo Reports* 2(9), 770-774.

Bogosian, G., Morris, P. J. L., and O'Neil, J. P. (1998). A mixed culture recovery method indicates that enteric bacteria do not enter the viable but nonculturable state. *Applied and Environmental Microbiology* 64(5), 1736-1742.

Bogosian, G., Sammons, L. E., Morris, P. J. L., ONeil, J. P., Heitkamp, M. A., and Weber, D. B. (1996). Death of the Escherichia coli K-12 strain W3110 in soil and water. *Applied and Environmental Microbiology* 62(11), 4114-4120.

Boucher, S. N., Slater, E. R., Chamberlain, A. H. L., and Adams, M. R. (1994). Production and Viability of Coccoid Forms of Campylobacter-Jejuni. *Journal of Applied Bacteriology* 77(3), 303-307.

Bouma, M. J., and Pascual, M. (2001). Seasonal and interannual cycles of endemic cholera in Bengal 1891-1940 in relation to climate and geography. *Hydrobiologia* 460, 147-156.

Bourassa, L., and Camilli, A. (2009). Glycogen contributes to the environmental persistence and transmission of Vibrio cholerae. *Molecular Microbiology* 72(1), 124-138.

Bovill, R. A., and Mackey, B. M. (1997). Resuscitation of 'non-culturable' cells from aged cultures of Campylobacter jejuni. *Microbiology-Uk* 143, 1575-1581.

Broza, M., Gancz, H., Halpern, M., and Kashi, Y. (2005). Adult non-biting midges: possible windborne carriers of Vibrio cholerae non-O1 non-O139. *Environmental Microbiology* 7(4), 576-585.

Broza, M., Gancz, H., and Kashi, Y. (2008). The association between non-biting midges and Vibrio cholerae. *Environmental Microbiology* 10(12), 3193-3200.

Cappelier, J. M., Besnard, V., Roche, S., Garrec, N., Zundel, E., Velge, P., and Federighi, M. (2005). Avirulence of viable but non-culturable Listeria monocytogenes cells demonstrated by *in vitro* and *in vivo* models. *Veterinary Research* 36(4), 589-599.

Cappelier, J. M., Besnard, V., Roche, S. M., Velge, P., and Federighi, M. (2007). Avirulent viable but non culturable cells of Listeria monocytogenes need the presence of an embryo to be recovered in egg yolk and regain virulence after recovery. *Veterinary Research* 38(4), 573-583.

Cappelier, J. M., Magras, C., Jouve, J. L., and Federighi, M. (1999a). Recovery of viable but non-culturable Campylobacter jejuni cells in two animal models. *Food Microbiology* 16(4), 375-383.

Cappelier, J. M., Minet, J., Magras, C., Colwell, R. R., and Federighi, M. (1999b). Recovery in embryonated eggs of viable but nonculturable Campylobacter jejuni cells and maintenance of ability to adhere to HeLa cells after resuscitation. *Applied and Environmental Microbiology* 65(11), 5154-5157.

Caro, A., Got, P., Lesne, J., Binard, S., and Baleux, B. (1999). Viability and virulence of experimentally stressed nonculturable Salmonella typhimurium. *Appl Environ Microbiol* 65(7), 3229-32.

Chaiyanan, S., Chaiyanan, S., Grim, C., Maugel, T., Huq, A., and Colwell, R. R. (2007). Ultrastructure of coccoid viable but non-culturable Vibrio cholerae. *Environmental Microbiology* 9(2), 393-402.

Chaiyanan, S., Chaiyanan, S., Huq, A., Maugel, T., and Colwell, R. R. (2001). Viability of the Nonculturable Vibrio cholerae O1 and O139. *Systematic and Applied Microbiology* 24(3), 331-341.

Chaveerach, P., ter Huurne, A. A. H. M., Lipman, L. J. A., and van Knapen, F. (2003). Survival and resuscitation of ten strains of Campylobacter jejuni and Campylobacter coli under acid conditions. *Applied and Environmental Microbiology* 69(1), 711-714.

Chen, C. H., Shimada, T., Elhadi, N., Radu, S., and Nishibuchi, M. (2004). Phenotypic and genotypic characteristics and epidemiological significance of ctx(+) strains of Vibrio cholerae isolated from seafood in Malaysia. *Applied and Environmental Microbiology* 70(4), 1964-1972.

Chowdhury, M. A. R., Huq, A., Xu, B., Madeira, F. J. B., and Colwell, R. R. (1997). Effect of alum on free-living and copepod-associated Vibrio cholerae O1 and O139. *Applied and Environmental Microbiology* 63(8), 3323-3326.

Chowdhury, M. A. R., Ravel, J., Hill, R. T., Huq, A. and Colwell, R. R. (1994). Physiology and molecular genetics of viable but non-culturable microorganisms. *In* "Biotechnology risk assessment: USEPA/USDA/Environment Canada: risk assessment methodologies" (M. Levin, Grim, C. and Scott, J. S., Ed.).

Chun, J., Grim, C. J., Hasan, N. A., Lee, J. H., Choi, S. Y., Haley, B. J., Taviani, E., Jeon, Y. S., Kim, D. W., Lee, J. H., Brettin, T. S., Bruce, D. C., Challacombe, J. F., Detter, J. C., Han, C. S., Munk, A. C., Chertkov, O., Meincke, L., Saunders, E., Walters, R. A., Huq, A., Nair, G. B., and Colwell, R. R. (2009). Comparative genomics reveals mechanism for short-term and long-term clonal transitions in pandemic Vibrio cholerae. *Proceedings of the National Academy of Sciences of the United States of America* 106(36), 15442-15447.

Colwell, R. R. (1996). Global climate and infectious disease: the cholera paradigm. *Science* 274 (5295), 2025-2031

Colwell, R. R., Brayton, P., Herrington, D., Tall, B., Huq, A., and Levine, M. M. (1996). Viable but non culturable Vibrio cholerae o1 revert to a cultivable state in the human intestine. *World Journal of Microbiology & Biotechnology* 12(1), 28-31.

Colwell, R. R., Seidler, R. J., Kaper, J., Joseph, S. W., Garges, S., Lockman, H., Maneval, D., Bradford, H., Roberts, N., Remmers, E., Huq, I., and Huq, A. (1981). Occurrence of Vibrio-Cholerae Serotype-O1 in Maryland and Louisiana Estuaries. *Applied and Environmental Microbiology* 41(2), 555-558.

Coutard, F., Crassous, P., Droguet, M., Gobin, E., Colwell, R. R., Pommepuy, M., and Hervio-Heath, D. (2007). Recovery in culture of viable but nonculturable Vibrio parahaemolyticus: regrowth or resuscitation? *Isme Journal* 1(2), 111-120.

Curras, M., Magarinos, B., Toranzo, A. E., and Romalde, J. L. (2002). Dormancy as a survival strategy of the fish pathogen Streptococcus parauberis in the marine environment. *Diseases of Aquatic Organisms* 52(2), 129-136.

Daniels, N. A., and Shafaie, A. (2000). A review of pathogenic Vibrio infections for clinicians. *Infections in Medicine* 17(10), 665-+.

Davis, J. W., and Sizemore, R. K. (1982). Incidence of Vibrio Species Associated with Blue Crabs (Callinectes-Sapidus) Collected from Galveston Bay, Texas. *Applied and Environmental Microbiology* 43(5), 1092-1097.

Day, A. P., and Oliver, J. D. (2004). Changes in membrane fatty acid composition during entry of Vibrio vulnificus into the viable but nonculturable state. *Journal of Microbiology* 42(2), 69-73.

del Campo, R., Russi, P., Mara, P., Mara, H., Peyrou, M., de Leon, I. P., and Gaggero, C. (2009). Xanthomonas axonopodis pv. citri enters the VBNC state after copper treatment and retains its virulence. *Fems Microbiology Letters* 298(2), 143-148.

Depaola, A., Capers, G. M., and Alexander, D. (1994). Densities of Vibrio-Vulnificus in the Intestines of Fish from the Us Gulf-Coast. *Applied and Environmental Microbiology* 60(3), 984-988.

DePaola, A., Nordstrom, J. L., Bowers, J. C., Wells, J. G., and Cook, D. W. (2003). Seasonal abundance of total and pathogenic Vibrio parahaemolyticus in Alabama oysters. *Applied and Environmental Microbiology* 69(3), 1521-1526.

Desnues, B., Cuny, C., Gregori, G., Dukan, S., Aguilaniu, H., and Nystrom, T. (2003). Differential oxidative damage and expression of stress defence regulons in culturable and non-culturable Escherichia coli cells. *EMBO Rep* 4(4), 400-4.

Diaper, J. P., and Edwards, C. (1994). Survival of Staphylococcus-Aureus in Lakewater Monitored by Flow-Cytometry. *Microbiology-Uk* 140, 35-42.

Dreux, N., Albagnac, C., Federighi, M., Carlin, F., Morris, C. E., and Nguyen-the, C. (2007). Viable but non-culturable Listeria monocytogenes on parsley leaves and absence of recovery to a culturable state. *Journal of Applied Microbiology* 103(4), 1272-1281.

Du, M., Chen, J., Zhang, X., Li, A., and Li, Y. (2007a). Characterization and resuscitation of viable but nonculturable Vibrio alginolyticus VIB283. *Arch Microbiol* 188(3), 283-8.

Du, M., Chen, J., Zhang, X., Li, A., Li, Y., and Wang, Y. (2007b). Retention of virulence in a viable but nonculturable Edwardsiella tarda isolate. *Appl Environ Microbiol* 73(4), 1349-54.

Dukan, S., Levi, Y., and Touati, D. (1997). Recovery of culturability of an HOCl-stressed population of Escherichia coli after incubation in phosphate buffer: resuscitation or regrowth? *Appl Environ Microbiol* 63(11), 4204-9.

Dusserre, E., Ginevra, C., Hallier-Soulier, S., Vandenesch, F., Festoc, G., Etienne, J., Jarraud, S., and Molmeret, M. (2008). A PCR-Based method for monitoring Legionella pneumophila in water samples detects viable but noncultivable legionellae that can recover their cultivability. *Applied and Environmental Microbiology* 74(15), 4817-4824.

Effendi, I., and Austin, B. (1995). Dormant Unculturable Cells of the Fish Pathogen Aeromonas-Salmonicida. *Microbial Ecology* 30(2), 183-192.

Eguchi, M., Fujiwara, E., and Miyamoto, N. (2000). Survival of Vibrio anguillarum in freshwater environments: adaptation or debilitation? *J Infect Chemother* 6(2), 126-9.

Emch, M., Feldacker, C., Islam, M. S., and Ali, M. (2008). Seasonality of cholera from 1974 to 2005: a review of global patterns. *International Journal of Health Geographics* 7, -.

Enroth, H., Wreiber, K., Rigo, R., Risberg, D., Uribe, A., and Engstrand, L. (1999). *In vitro* aging of Helicobacter pylori: Changes in morphology, intracellular composition and surface properties. *Helicobacter* 4(1), 7-16.

Falcioni, T., Papa, S., Campana, R., Manti, A., Battistelli, M., and Baffone, W. (2008). State transitions of Vibrio parahaemolyticus VBNC cells evaluated by flow cytometry. *Cytometry Part B-Clinical Cytometry* 74B(5), 272-281.

Faruque, S. M., Bin Naser, I., Islam, M. J., Faruque, A. S. G., Ghosh, A. N., Nair, G. B., Sack, D. A., and Mekalanos, J. J. (2005). Seasonal epidemics of cholera inversely correlate with the prevalence of environmental cholera phages. *Proceedings of the National Academy of Sciences of the United States of America* 102(5), 1702-1707.

Faruque, S. M., Biswas, K., Udden, S. M. N., Ahmad, Q. S., Sack, D. A., Nair, G. B., and Mekalanos, J. J. (2006a). Transmissibility of cholera: *In vivo*-formed biofilms and their relationship to infectivity and persistence in the environment. *Proceedings of the National Academy of Sciences of the United States of America* 103(16), 6350-6355.

Faruque, S. M., Islam, M. J., Ahmad, Q. S., Biswas, K., Faruque, A. S. G., Nair, G. B., Sack, R. B., Sack, D. A., and Mekalanos, J. J. (2006b). An improved technique for isolation of environmental Vibrio cholerae with epidemic potential: Monitoring the emergence of a multiple-antibiotic-resistant epidemic strain in Bangladesh. *Journal of Infectious Diseases* 193(7), 1029-1036.

Forsman, M., Henningson, E. W., Larsson, E., Johansson, T., and Sandstrom, G. (2000). Francisella tularensis does not manifest virulence in viable but non-culturable state. *FEMS Microbiol Ecol* 31(3), 217-224.

Forssman, B., Mannes, T., Musto, J., Gottlieb, T., Robertson, G., Natoli, J. D., Shadbolt, C., Biffin, B., and Gupta, L. (2007). Vibrio cholerae O1 El Tor cluster in Sydney linked to imported whitebait. *Medical Journal of Australia* 187(6), 345-347.

Gaffga, N. H., Tauxe, R. V., and Mintz, E. D. (2007). Cholera: A new homeland in Africa? *American Journal of Tropical Medicine and Hygiene* 77(4), 705-713.

Ghezzi, J. I., and Steck, T. R. (1999). Induction of the viable but non-culturable condition in Xanthomonas campestris pv. campestris in liquid microcosms and sterile soil. *Fems Microbiology Ecology* 30(3), 203-208.

Gil, A. I., Louis, V. R., Rivera, I. N. G., Lipp, E., Huq, A., Lanata, C. F., Taylor, D. N., Russek-Cohen, E., Choopun, N., Sack, R. B., and Colwell, R. R. (2004). Occurrence and distribution of Vibrio cholerae in the coastal environment of Peru. *Environmental Microbiology* 6(7), 699-706.

Greenstone, G. (2009). A commentary on cholera: The scourge that never dies. *BCMJ* 51(4), 164-167.

Grey, B. E., and Steck, T. R. (2001a). The viable but nonculturable state of Ralstonia solanacearum may be involved in long-term survival and plant infection. *Applied and Environmental Microbiology* 67(9), 3866-3872.

Grey, B. N., and Steck, T. R. (2001b). Concentrations of copper thought to be toxic to Escherichia coli can induce the viable but nonculturable condition. *Applied and Environmental Microbiology* 67(11), 5325-5327.

Grim, C. J., Zo, Y. G., Hasan, N. A., Ali, A., Chowdhury, W. B., Islam, A., Rashid, M. H., Alam, M., Morris, J. G., Huq, A., and Colwell, R. R. (2009). RNA Colony Blot Hybridization Method for Enumeration of Culturable Vibrio cholerae and Vibrio mimicus Bacteria. *Applied and Environmental Microbiology* 75(17), 5439-5444.

Guillou, S., Leguerinel, I., Garrec, N., Renard, M. A., Cappelier, J. M., and Federighi, M. (2008). Survival of Campylobacter jejuni in mineral bottled water according to difference in mineral content: Application of the Weibull model. *Water Research* 42(8-9), 2213-2219.

Gupte, A. R., de Rezende, C. L. E., and Joseph, S. W. (2003). Induction and resuscitation of viable but nonculturable Salmonella enterica serovar typhimurium DT104. *Applied and Environmental Microbiology* 69(11), 6669-6675.

Hald, B., Knudsen, K., Lind, P., and Madsen, M. (2001). Study of the infectivity of saline-stored Campylobacter jejuni for day-old chicks. *Appl Environ Microbiol* 67(5), 2388-92.

Halpern, A., Broza, Y. B., Mittler, S., Arakawa, E., and Broza, A. (2004). Chironomid egg masses as a natural reservoir of Vibrio cholerae non-O1 and non-O139 in freshwater habitats (vol 47, pg 341, 2003). *Microbial Ecology* 48(2), 285-285.

Halpern, M., Landsberg, O., Raats, D., and Rosenberg, E. (2007). Culturable and VBNC Vibrio cholerae: Interactions with chironomid egg masses and their bacterial population. *Microbial Ecology* 53(2), 285-293.

Halpern, M., Senderovich, Y., and Izhaki, I. (2008). Waterfowl-The Missing Link in Epidemic and Pandemic Cholera Dissemination? *Plos Pathogens* 4(10), -.

Heidelberg, J. F., Shahamat, M., Levin, M., Rahman, I., Stelma, G., Grim, C., and Colwell, R. R. (1997). Effect of aerosolization on culturability and viability of gram-negative bacteria. *Applied and Environmental Microbiology* 63(9), 3585-3588.

Heim, S., Lleo, M. D., Bonato, B., Guzman, C. A., and Canepari, P. (2002). The viable but nonculturable state and starvation are different stress responses of Enterococcus faecalis, as determined by proteome analysis. *Journal of Bacteriology* 184(23), 6739-6745.

Hlady, W, G. and Klontz, K. C. (1996). The epidemiology of Vibrio infections in Florida, 1981-1993. *J Infect Dis* 173(5)1176-1183.

Hoi, L., Larsen, J. L., Dalsgaard, I., and Dalsgaard, A. (1998). Occurrence of Vibrio vulnificus biotypes in Danish marine environments. *Applied and Environmental Microbiology* 64(1), 7-13.

Hood, M. A., and WInter, P. A. (1997). Attachment of Vibrio cholerae under various environmental conditions and to selected substrates. *Fems Microbiology Ecology* 22(3), 215-223.

Huq, A., Colwell, R. R., Rahman, R., Ali, A., Chowdhury, M. A. R., Parveen, S., Sack, D. A., and Russekcohen, E. (1990). Detection of Vibrio-Cholerae O1 in the Aquatic Environment by Fluorescent-Monoclonal Antibody and Culture Methods. *Applied and Environmental Microbiology* 56(8), 2370-2373.

Huq, A., Sack, R. B., Nizam, A., Longini, I. M., Nair, G. B., Ali, A., Morris, J. G., Khan, M. N. H., Siddique, A. K., Yunus, M., Albert, M. J., Sack, D. A., and Colwell, R. R. (2005). Critical factors influencing the occurrence of Vibrio cholerae in the environment of Bangladesh. *Applied and Environmental Microbiology* 71(8), 4645-4654.

Huq, A., Xu, B., Chowdhury, M. A. R., Islam, M. S., Montilla, R., and Colwell, R. R. (1996). A simple filtration method to remove plankton-associated Vibrio cholerae in raw water supplies in developing countries. *Applied and Environmental Microbiology* 62(7), 2508-2512.

Hwang, M. G., Katayama, H., and Ohgaki, S. (2006). Effect of intracellular resuscitation of Legionella pneumophila in Acanthamoeba polyphage cells on the antimicrobial properties of silver and copper. *Environmental Science & Technology* 40(23), 7434-7439.

Islam, M. S., Alam, M. J., Begum, A., Rahim, Z., Felsenstein, A., and Albert, M. J. (1996). Occurrence of culturable Vibrio cholerae O139 with ctx gene in various components of the aquatic environment in Bangladesh. *Transactions of the Royal Society of Tropical Medicine and Hygiene* 90(2), 128-128.

Islam, M. S., Drasar, B. S., and Bradley, D. J. (1990). Survival of Toxigenic Vibrio-Cholerae O1 with a Common Duckweed, Lemna-Minor, in Artificial Aquatic Ecosystems. *Transactions of the Royal Society of Tropical Medicine and Hygiene* 84(3), 422-424.

Islam, M. S., Drasar, B. S., and Sack, R. B. (1993). The Aquatic Environment as a Reservoir of Vibrio-Cholerae - a Review. *Journal of Diarrhoeal Diseases Research* 11(4), 197-206.

Islam, M. S., Rahim, Z., Alam, M. J., Begum, S., Moniruzzaman, S. M., Umeda, A., Amako, K., Albert, M. J., Sack, R. B., Huq, A., and Colwell, R. R. (1999). Association of Vibrio cholerae O1 with the cyanobacterium, Anabaena sp., elucidated by polymerase chain reaction and transmission electron microscopy. *Transactions of the Royal Society of Tropical Medicine and Hygiene* 93(1), 36-40.

Islam, M. S., Sharker, M. A. Y., Rheman, S., Hossain, S., Mahmud, Z. H., Islam, M. S., Uddin, A. M. K., Yunus, M., Osman, M. S., Ernst, R., Rector, I., Larson, C. P., Luby, S. P., Endtz, H. P., and Cravioto, A. (2009). Effects of local climate variability on transmission dynamics of cholera in Matlab, Bangladesh. *Transactions of the Royal Society of Tropical Medicine and Hygiene* 103(11), 1165-1170.

Islam, M. S., Talukder, K. A., Khan, N. H., Mahmud, Z. H., Rahman, M. Z., Nair, G. B., Siddique, A. K. M., Yunus, M., Sack, D. A., Sack, R. B., Huq, A., and Colwell, R. R. (2004). Variation of toxigenic Vibrio cholerae o1 in the aquatic environment of Bangladesh and its correlation with the clinical strains. *Microbiology and Immunology* 48(10), 773-777.

Israely, T., Banin, E., and Rosenberg, E. (2001). Growth, differentiation and death of Vibrio shiloi in coral tissue as a function of seawater temperature. *Aquatic Microbial Ecology* 24(1), 1-8.

Jiang, X. P., and Chai, T. J. (1996). Survival of Vibrio parahaemolyticus at low temperatures under starvation conditions and subsequent resuscitation of viable, nonculturable cells. *Applied and Environmental Microbiology* 62(4), 1300-1305.

Kaneko, T., and Colwell, R. R. (1973). Ecology of Vibrio-Parahaemolyticus in Chesapeake Bay. *Journal of Bacteriology* 113(1), 24-32.

Kaneko, T., and Colwell, R. R. (1975). Incidence of Vibrio-Parahaemolyticus in Chesapeake Bay. *Applied Microbiology* 30(2), 251-257.

Kaper, J., Lockman, H., Colwell, R. R., and Joseph, S. W. (1979). Ecology, Serology, and Enterotoxin Production of Vibrio-Cholerae in Chesapeake Bay. *Applied and Environmental Microbiology* 37(1), 91-103.

Kaper, J. B., Morris, J. G., and Levine, M. M. (1995). Cholera. *Clinical Microbiology Reviews* 8(1), 48-86.

Karaolis, D. K. R., Lan, R. T., and Reeves, P. R. (1995). The 6th and 7th Cholera Pandemics Are Due to Independent Clones Separately Derived from Environmental, Nontoxigenic, Non-O1 Vibrio-Cholerae. *Journal of Bacteriology* 177(11), 3191-3198.

Kaspar, C. W., and Tamplin, M. L. (1993). Effects of Temperature and Salinity on the Survival of Vibrio-Vulnificus in Seawater and Shellfish. *Applied and Environmental Microbiology* 59(8), 2425-2429.

Kaysner, C. A., Tamplin, M. L., Wekell, M. M., Stott, R. F., and Colburn, K. G. (1989). Survival of Vibrio-Vulnificus in Shellstock and Shucked Oysters (Crassostrea-Gigas and Crassostrea-Virginica) and Effects of Isolation Medium on Recovery. *Applied and Environmental Microbiology* 55(12), 3072-3079.

Kiehl, W. (1998). Cases of cholera imported to Germany from Kenya. *Euro Surveill* 2(6), 1261.

King, A. A., Ionides, E. L., Pascual, M., and Bouma, M. J. (2008). Inapparent infections and cholera dynamics. *Nature* 454(7206), 877-U29.

Kirschner, A. K. T., Schlesinger, J., Farnleitner, A. H., Hornek, R., Suss, B., Golda, B., Herzig, A., and Reitner, B. (2008). Rapid growth of planktonic Vibrio cholerae Non-O1/Non-O139 strains in a large alkaline lake in Austria: Dependence on temperature and dissolved organic carbon quality. *Applied and Environmental Microbiology* 74(7), 2004-2015.

Klancnik, A., Guzej, B., Jamnik, P., Vuckovic, D., Abram, M., and Mozina, S. S. (2009). Stress response and pathogenic potential of Campylobacter jejuni cells exposed to starvation. *Research in Microbiology* 160(5), 345-352.

Koechlein, D. J., and Krieg, N. R. (1998). Viable but nonculturable coccoid forms of Prolinoborus fasciculus (Aquaspirillum fasciculus). *Canadian Journal of Microbiology* 44(9), 910-912.

Koelle, K., Rodo, X., Pascual, M., Yunus, M., and Mostafa, G. (2005). Refractory periods and climate forcing in cholera dynamics. *Nature* 436(7051), 696-700.

Kolling, G. L., and Matthews, K. R. (2001). Examination of recovery *in vitro* and *in vivo* of nonculturable Escherichia coli O157 : H7. *Applied and Environmental Microbiology* 67(9), 3928-+.

Kushmaro, A., Banin, E., Loya, Y., Stackebrandt, E., and Rosenberg, E. (2001). Vibrio shiloi sp nov., the causative agent of bleaching of the coral Oculina patagonica. *International Journal of Systematic and Evolutionary Microbiology* 51, 1383-1388.

Leal, N. C., Figueiroa, A. C. T., Cavalcanti, V. O., da Silva, S. C., Leal-Balbino, T. C., de Almeida, A. M. P., and Hofer, E. (2008). Characterization of Vibrio cholerae isolated from the aquatic basins of the State of Pernambuco, Brazil. *Transactions of the Royal Society of Tropical Medicine and Hygiene* 102(3), 272-276.

Lebaron, P., and Joux, F. (1994). Flow Cytometric Analysis of the Cellular DNA Content of Salmonella-Typhimurium and Alteromonas-Haloplanktis during Starvation and Recovery in Seawater. *Applied and Environmental Microbiology* 60(12), 4345-4350.

Lee, J. L., and Levin, R. E. (2009). A comparative study of the ability of EMA and PMA to distinguish viable from heat killed mixed bacterial flora from fish fillets. *Journal of Microbiological Methods* 76(1), 93-96.

Lee, J. V., Bashford, D. J., Donovan, T. J., Furniss, A. L., and West, P. A. (1982). The Incidence of Vibrio-Cholerae in Water, Animals and Birds in Kent, England. *Journal of Applied Bacteriology* 52(2), 281-291.

Li, J., Kolling, G. L., Matthews, K. R., and Chikindas, M. L. (2003). Cold and carbon dioxide used as multi-hurdle preservation do not induce appearance of viable but non-culturable Listeria monocytogenes. *J Appl Microbiol* 94(1), 48-53.

Lindback, T., Rottenberg, M. E., Roche, S. M., and Rorvik, L. M. (2010). The ability to enter into an avirulent viable but non-culturable (VBNC) form is widespread among Listeria monocytogenes isolates from salmon, patients and environment. *Veterinary Research* 41(1), -.

Linder, K., and Oliver, J. D. (1989). Membrane Fatty-Acid and Virulence Changes in the Viable but Nonculturable State of Vibrio-Vulnificus. *Applied and Environmental Microbiology* 55(11), 2837-2842.

Lipp, E. K., Huq, A., and Colwell, R. R. (2002). Effects of global climate on infectious disease: the cholera model. *Clinical Microbiology Reviews* 15(4), 757-+.

Lipp, E. K., Rivera, I. N. G., Gil, A. I., Espeland, E. M., Choopun, N., Louis, V. R., Russek-Cohen, E., Huq, A., and Colwell, R. R. (2003). Direct detection of Vibrio cholerae and ctxA in Peruvian coastal water and plankton by PCR. *Applied and Environmental Microbiology* 69(6), 3676-3680.

Lleo, M. D., Pierobon, S., Tafi, M. C., Signoretto, C., and Canepari, P. (2000). mRNA detection by reverse transcription-PCR for monitoring viability over time in an Enterococcus faecalis viable but nonculturable population maintained in a laboratory microcosm. *Applied and Environmental Microbiology* 66(10), 4564-4567.

Lleo, M. M., Bonato, B., Tafi, M. C., Signoretto, C., Boaretti, M., and Canepari, P. (2001). Resuscitation rate in different enterococcal species in the viable but non-culturable state. *Journal of Applied Microbiology* 91(6), 1095-1102.

Lothigius, A., Sjoling, A., Svennerholm, A. M., and Bolin, I. (2010). Survival and gene expression of enterotoxigenic Escherichia coli during long-term incubation in sea water and freshwater. *Journal of Applied Microbiology* 108(4), 1441-1449.

Louis, V. R., Russek-Cohen, E., Choopun, N., Rivera, I. N. G., Gangle, B., Jiang, S. C., Rubin, A., Patz, J. A., Huq, A., and Colwell, R. R. (2003). Predictability of Vibrio cholerae in Chesapeake Bay. *Applied and Environmental Microbiology* 69(5), 2773-2785.

Maalej, S., Denis, M., and Dukan, S. (2004). Temperature and growth-phase effects on Aeromonas hydrophila survival in natural seawater microcosms: role of protein synthesis and nucleic acid content on viable but temporarily nonculturable response. *Microbiology-Sgm* 150, 181-187.

Mahale, D. P., Khade, R. G. and Vaidya, V. K. (2008). Microbiological analysis of street vended fruit juices from Mumbai city, India. *Internet Journal of Food Safety* 10, 31-34.

Mahmud, Z. H., Kassu, A., Mohammad, A., Yamato, M., Bhuiyan, N. A., Nair, G. B., and Ota, F. (2006). Isolation and molecular characterization of toxigenic Vibrio parahaemolyticus from the Kii Channel, Japan. *Microbiological Research* 161(1), 25-37.

Marco-Noales, E., Biosca, E. G., and Amaro, C. (1999). Effects of salinity and temperature on long-term survival of the eel pathogen Vibrio vulnificus biotype 2 (serovar E). *Applied and Environmental Microbiology* 65(3), 1117-1126.

Marco-Noales, E., Biosca, E. G., Rojo, C., and Amaro, C. (2004). Influence of aquatic microbiota on the survival in water of the human and eel pathogen Vibrio vulnificus serovar E. *Environmental Microbiology* 6(4), 364-376.

Martinez-Urtaza, J., Lozano-Leon, A., DePaola, A., Ishibashi, M., Shimada, K., Nishibuchi, M., and Liebana, E. (2004). Characterization of pathogenic Vibrio parahaemolyticus isolates from clinical sources in Spain and comparison with Asian and North American pandemic isolates. *Journal of Clinical Microbiology* 42(10), 4672-4678.

Matsumoto, C., Okuda, J., Ishibashi, M., Iwanaga, M., Garg, P., Rammamurthy, T., Wong, H. C., Depaola, A., Kim, Y. B., Albert, M. J., and Nishibuchi, M. (2000). Pandemic spread of an O3 : K6 clone of Vibrio parahaemolyticus and emergence of related strains evidenced by arbitrarily primed PCR and toxRS sequence analyses. *Journal of Clinical Microbiology* 38(2), 578-585.

Matz, C., McDougald, D., Moreno, A. M., Yung, P. Y., Yildiz, F. H., and Kjelleberg, S. (2005). Biofilm formation and phenotypic variation enhance predation-driven persistence of Vibrio cholerae. *Proceedings of the National Academy of Sciences of the United States of America* 102(46), 16819-16824.

Mccarthy, S. A., and Khambaty, F. M. (1994). International Dissemination of Epidemic Vibrio-Cholerae by Cargo Ship Ballast and Other Nonpotable Waters. *Applied and Environmental Microbiology* 60(7), 2597-2601.

McDougald, D., Rice, S. A., Weichart, D., and Kjelleberg, S. (1998). Nonculturability: adaptation or debilitation? *Fems Microbiology Ecology* 25(1), 1-9.

Medema, G. J., Schets, F. M., Vandegiessen, A. W., and Havelaar, A. H. (1992). Lack of Colonization of 1-Day-Old Chicks by Viable, Non-Culturable Campylobacter-Jejuni. *Journal of Applied Bacteriology* 72(6), 512-516.

Meibom, K. L., Li, X. B. B., Nielsen, A. T., Wu, C. Y., Roseman, S., and Schoolnik, G. K. (2004). The Vibrio cholerae chitin utilization program. *Proceedings of the National Academy of Sciences of the United States of America* 101(8), 2524-2529.

Merrell, D. S., Butler, S. M., Qadri, F., Dolganov, N. A., Alam, A., Cohen, M. B., Calderwood, S. B., Schoolnik, G. K., and Camilli, A. (2002). Host-induced epidemic spread of the cholera bacterium. *Nature* 417(6889), 642-645.

Minami, A., Hashimoto, S., Abe, H., Arita, M., Taniguchi, T., Honda, T., Miwatani, T., and Nishibuchi, M. (1991). Cholera Enterotoxin Production in Vibrio-Cholerae-O1 Strains Isolated from the Environment and from Humans in Japan. *Applied and Environmental Microbiology* 57(8), 2152-2157.

Mizunoe, Y., Wai, S. N., Takade, A., and Yoshida, S. (1999). Restoration of culturability of starvation-stressed and low-temperature-stressed Escherichia coli O157 cells by using H2O2-degrading compounds. *Archives of Microbiology* 172(1), 63-67.

Mourino-Perez, R. R. (1998). Oceanography and the seventh cholera pandemic. *Epidemiology* 9(3), 355-357.

Muela, A., Arana, I. I., Justo, J. I., Seco, C., and Barcina, I. I. (1999). Changes in DNA Content and Cellular Death during a Starvation-Survival Process of Escherichia coli in River Water. *Microb Ecol* 37(1), 62-69.

Muela, A., Seco, C., Camafeita, E., Arana, I., Orruno, M., Lopez, J. A., and Barcina, I. (2008). Changes in Escherichia coli outer membrane subproteome under environmental conditions inducing the viable but nonculturable state. *Fems Microbiology Ecology* 64(1), 28-36.

Mukamolova, G. V., Turapov, O., Malkin, J., Woltmann, G., and Barer, M. R. (2010). Resuscitation-promoting Factors Reveal an Occult Population of Tubercle Bacilli in Sputum. *American Journal of Respiratory and Critical Care Medicine* 181(2), 174-180.

Mukamolova, G. V., Yanopolskaya, N. D., Votyakova, T. V., Popov, V. I., Kaprelyants, A. S., and Kell, D. B. (1995). Biochemical-Changes Accompanying the Long-Term Starvation of Micrococcus-Luteus Cells in Spent Growth-Medium. *Archives of Microbiology* 163(5), 373-379.

Nair, G. B., Ramamurthy, T., Bhattacharya, S. K., Dutta, B., Takeda, Y., and Sack, D. A. (2007). Global dissemination of Vibrio parahaemolyticus serotype O3 : K6 and its serovariants. *Clinical Microbiology Reviews* 20(1), 39-+.

Nair, G. B., Sarkar, B. L., De, S. P., Chakrabarti, M. K., Bhadra, R. K., and Pal, S. C. (1988). Ecology of Vibrio-Cholerae in the Fresh-Water Environs of Calcutta, India. *Microbial Ecology* 15(2), 203-215.

Nelson, E. J., Chowdhury, A., Flynn, J., Schild, S., Bourassa, L., Shao, Y., LaRocque, R. C., Calderwood, S. B., Qadri, F., and Camilli, A. (2008). Transmission of Vibrio cholerae Is Antagonized by Lytic Phage and Entry into the Aquatic Environment. *Plos Pathogens* 4(10), -.

Nilsson, L., Oliver, J. D., and Kjelleberg, S. (1991). Resuscitation of Vibrio-Vulnificus from the Viable but Nonculturable State. *Journal of Bacteriology* 173(16), 5054-5059.

Normanno, G., Parisi, A., Addante, N., Quaglia, N. C., Dambrosio, A., Montagna, C., and Chiocco, D. (2006). Vibrio parahaemolyticus, Vibrio vulnificus and microorganisms of fecal origin in mussels (Mytilus galloprovincialis) sold in the Puglia region (Italy). *Int J Food Microbiol* 106(2), 219-22.

Nusrin, S., Gil, A. I., Bhuiyan, N. A., Safa, A., Asakura, M., Lanata, C. F., Hall, E., Miranda, H., Huapaya, B., Vargas, G. C., Luna, M. A., Sack, D. A., Yamasaki, S., and Nair, G. B. (2009). Peruvian Vibrio cholerae O1 El Tor strains possess a distinct region in the Vibrio seventh pandemic island-II that differentiates them from the prototype seventh pandemic El Tor strains. *Journal of Medical Microbiology* 58(3), 342-354.

Nystrom, T. (2001). Not quite dead enough: on bacterial life, culturability, senescence, and death. *Archives of Microbiology* 176(3), 159-164.

O'Brien, S. a. S.-S., R. (1998). Cholera in tourists returning from Kenya. *Euro Surveill* 2(16), 1229.

Ogg, J. E., Ryder, R. A., and Smith, H. L. (1989). Isolation of Vibrio-Cholerae from Aquatic Birds in Colorado and Utah. *Applied and Environmental Microbiology* 55(1), 95-99.

Olafsen, J. A., Mikkelsen, H. V., Glaever, H. M., and Hansen, G. H. (1993). Indigenous Bacteria in Hemolymph and Tissues of Marine Bivalves at Low-Temperatures. *Applied and Environmental Microbiology* 59(6), 1848-1854.

Oliver, J. D. (2005). The viable but nonculturable state in bacteria. *Journal of Microbiology* 43, 93-100.

Oliver, J. D., and Bockian, R. (1995). In-Vivo Resuscitation, and Virulence Towards Mice, of Viable but Nonculturable Cells of Vibrio-Vulnificus. *Applied and Environmental Microbiology* 61(7), 2620-2623.

Oliver, J. D., Dagher, M., and Linden, K. (2005). Induction of Escherichia coli and Salmonella typhimurium into the viable but nonculturable state following chlorination of wastewater. *J Water Health* 3(3), 249-57.

Oliver, J. D., Hite, F., Mcdougald, D., Andon, N. L., and Simpson, L. M. (1995). Entry into, and Resuscitation from, the Viable but Nonculturable State by Vibrio-Vulnificus in an Estuarine Environment. *Applied and Environmental Microbiology* 61(7), 2624-2630.

Oliver, J. D., Nilsson, L., and Kjelleberg, S. (1991). Formation of Nonculturable Vibrio-Vulnificus Cells and Its Relationship to the Starvation State. *Applied and Environmental Microbiology* 57(9), 2640-2644.

Olivier, V., Salzman, N. H., and Satchell, K. J. F. (2007). Prolonged colonization of mice by Vibrio cholerae El tor O1 depends on accessory toxins. *Infection and Immunity* 75(10), 5043-5051.

Ordax, M., Marco-Noales, E., Lopez, M. M., and Biosca, E. G. (2006). Survival strategy of Erwinia amylovora against copper: Induction of the viable-but-nonculturable state. *Applied and Environmental Microbiology* 72(5), 3482-3488.

Ottaviani, D., Leoni, F., Rocchegiani, E., Santarelli, S., Masini, L., Di Trani, V., Canonico, C., Pianetti, A., Tega, L., and Carraturo, A. (2009). Prevalence and virulence properties of non-O1 non-O139 Vibrio cholerae strains from seafood and clinical samples collected in Italy. *International Journal of Food Microbiology* 132(1), 47-53.

Panutdaporn, N., Kawamoto, K., Asakura, H., and Makino, S. I. (2006). Resuscitation of the viable but non-culturable state of Salmonella enterica serovar Oranienburg by recombinant resuscitation-promoting factor derived from Salmonella Typhimurium strain LT2. *International Journal of Food Microbiology* 106(3), 241-247.

Pascual, M., Bouma, M. J., and Dobson, A. P. (2002). Cholera and climate: revisiting the quantitative evidence. *Microbes and Infection* 4(2), 237-245.

Paz, S., and Broza, M. (2007). Wind direction and its linkage with Vibrio cholerae dissemination. *Environmental Health Perspectives* 115(2), 195-200.

Pruzzo, C., Tarsi, R., Lleo, M. D., Signoretto, C., Zampini, M., Colwell, R. R., and Canepari, P. (2002). In vitro adhesion to human cells by viable but nonculturable Enterococcus faecalis. *Current Microbiology* 45(2), 105-110.

Pruzzo, C., Tarsi, R., Lleo, M. D., Signoretto, C., Zampini, M., Pane, L., Colwell, R. R., and Canepari, P. (2003). Persistence of adhesive properties in Vibrio cholerae after long-term exposure to sea water. *Environmental Microbiology* 5(10), 850-858.

Puglielli, L., Cattrini, C., Resa, J. J. G., Velasques, M., and Garcia, L. M. L. (1992). Symptomless Carriage of Vibrio-Cholerae in Peru. *Lancet* 339(8800), 1057-1057.

Quilici, M. L., Robert-Pillot, A., Picart, J., and Fournier, J. M. (2005). Pandemic Vibrio parahaemolyticus O3 : K6 spread, France. *Emerging Infectious Diseases* 11(7), 1148-1149.

Rahman, I., Shahamat, M., Chowdhury, M. A. R., and Colwell, R. R. (1996). Potential virulence of viable but nonculturable Shigella dysenteriae type 1. *Applied and Environmental Microbiology* 62(1), 115-120.

Rahman, I., Shahamat, M., Kirchman, P. A., Russekcohen, E., and Colwell, R. R. (1994). Methionine Uptake and Cytopathogenicity of Viable but Nonculturable Shigella-Dysenteriae Type-1. *Applied and Environmental Microbiology* 60(10), 3573-3578.

Rahman, M. H., Suzuki, S., and Kawai, K. (2001). Formation of viable but non-culturable state (VBNC) of Aeromonas hydrophila and its virulence in goldfish, Carassius auratus. *Microbiological Research* 156(1), 103-106.

Ramirez, G. D., Buck, G. W., Smith, A. K., Gordon, K. V., and Mott, J. B. (2009). Incidence of Vibrio vulnificus in estuarine waters of the south Texas Coastal Bend region. *Journal of Applied Microbiology* 107(6), 2047-2053.

Randa, M. A., Polz, M. F., and Lim, E. (2004). Effects of temperature and salinity on Vibrio vulnificus population dynamics as assessed by quantitative PCR. *Appl Environ Microbiol* 70(9), 5469-76.

Ravel, J., Knight, I. T., Monahan, C. E., Hill, R. T., and Colwell, R. R. (1995). Temperature-Induced Recovery of Vibrio-Cholerae from the Viable but Nonculturable State - Growth or Resuscitation. *Microbiology-Uk* 141, 377-383.

Rawlings, T. K., Ruiz, G. M., and Colwell, R. R. (2007). Association of Vibrio cholerae O1 El Tor and O139 Bengal with the copepods Acartia tonsa and Eurytemora affinis. *Applied and Environmental Microbiology* 73(24), 7926-7933.

Rice, S. A., McDougald, D. and Kjelleberg, S. (2000). *Vibrio vulnificus*: a physiological and genetic approach to the viable but nonculturable response. *J Infect Chemother* 6, 115-120.

Rollins, D. M., and Colwell, R. R. (1986). Viable but Nonculturable Stage of Campylobacter-Jejuni and Its Role in Survival in the Natural Aquatic Environment. *Applied and Environmental Microbiology* 52(3), 531-538.

Rosenberg, E., and Falkovitz, L. (2004). The Vibrio shiloi/Oculina patagonica model system of coral bleaching. *Annual Review of Microbiology* 58, 143-159.

Roszak, D. B., and Colwell, R. R. (1987). Survival Strategies of Bacteria in the Natural-Environment. *Microbiological Reviews* 51(3), 365-379.

Sack, R. B., Siddique, A. K., Longini, I. M., Nizam, A., Yunus, M., Islam, M. S., Morris, J. G., Ali, A., Huq, A., Nair, G. B., Qadri, F., Faruque, S. M., Sack, D. A., and Colwell, R. R. (2003). A 4-year study of the epidemiology of Vibrio cholerae in four rural areas of Bangladesh. *Journal of Infectious Diseases* 187(1), 96-101.

Sasaki, S., Suzuki, H., Fujino, Y., Kimura, Y., and Cheelo, M. (2009). Impact of Drainage Networks on Cholera Outbreaks in Lusaka, Zambia. *American Journal of Public Health* 99(11), 1982-1987.

Seas, C., Miranda, J., Gil, A. I., Leon-Barua, R., Patz, J., Huq, A., Colwell, R. R., and Sack, R. B. (2000). New insights on the emergence of cholera in Latin America during 1991: the Peruvian experience. *American Journal of Tropical Medicine and Hygiene* 62(4), 513-517.

Senderovich, Y., Izhaki, I., and Halpern, M. (2010). Fish as Reservoirs and Vectors of Vibrio cholerae. *Plos One* 5(1), -.

Sharma, C., Ghosh, A., Dalsgaard, A., Forslund, A., Ghosh, R. K., Bhattacharya, S. K., and Nair, G. B. (1998). Molecular evidence that a distinct Vibrio cholerae O1 biotype El Tor strain in Calcutta may have spread to the African continent. *Journal of Clinical Microbiology* 36(3), 843-844.

Signoretto, C., Burlacchini, G., Lleo, M. D., Pruzzo, C., Zampini, M., Pane, L., Franzini, G., and Canepari, P. (2004). Adhesion of Enterococcus faecalis in the nonculturable state to plankton is the main mechanism responsible for persistence of this bacterium in both lake and seawater. *Applied and Environmental Microbiology* 70(11), 6892-6896.

Signoretto, C., Burlacchini, G., Pruzzo, C., and Canepari, P. (2005). Persistence of *Enterococcus faecalis* in aquatic environments via surface interactions with copepods. *Applied and Environmental Microbiology* 71(5), 2756-2761.

Singleton, F. L., Attwell, R., Jangi, S., and Colwell, R. R. (1982). Effects of Temperature and Salinity on Vibrio-Cholerae Growth. *Applied and Environmental Microbiology* 44(5), 1047-1058.

Smith, R. J., Newton, A. T., Harwood, C. R., and Barer, M. R. (2002). Active but nonculturable cells of Salmonella enterica serovar Typhimurium do not infect or colonize mice. *Microbiology-Sgm* 148, 2717-2726.

Snoussi, M., Noumi, E., Usai, D., Sechi, L. A., Zanetti, S., and Bakhrouf, A. (2008). Distribution of some virulence related-properties of Vibrio alginolyticus strains isolated from Mediterranean seawater (Bay of Khenis, Tunisia): investigation of eight Vibrio cholerae virulence genes. *World Journal of Microbiology & Biotechnology* 24(10), 2133-2141.

Steinert, M., Emody, L., Amann, R., and Hacker, J. (1997). Resuscitation of viable but nonculturable Legionella pneumophila Philadelphia JR32 by Acanthamoeba castellanii. *Applied and Environmental Microbiology* 63(5), 2047-2053.

Stine, O. C., Alam, M., Tang, L., Nair, G. B., Siddique, A. K., Faruque, S. M., Huq, A., Colwell, R., Sack, R. B., and Morris, J. G. (2008). Seasonal cholera from multiple small outbreak, rural Bangladesh. *Emerging Infectious Diseases* 14(5), 831-833.

Strauss, R., Stirling, J., Hausmann, M., Wewalka, G., Thalhammer, F., Hrabcik, H. and Bruns, C. (2004). Two cases of imported cholera from India in Austrian travellers. *Euro Surveill* 8(30), 2507.

Strumbelj, I., Prelog, I., Kotar, T., Dovecar, D., Petras, T. and Socan, M. (2005). A case of Vibrio cholerae non-O1, non-O139 septicaemia in Slovenia, imported from Tunisia, July 2005. *Euro Surveill* 10(42), 2817.

Su, Y. C., and Liu, C. C. (2007). Vibrio parahaemolyticus: A concern of seafood safety. *Food Microbiology* 24(6), 549-558.

Sun, F., Chen, J., Zhong, L., Zhang, X. H., Wang, R., Guo, Q., and Dong, Y. (2008). Characterization and virulence retention of viable but nonculturable Vibrio harveyi. *FEMS Microbiol Ecol* 64(1), 37-44.

Sung, H. H., Chen, C. K., Shih, P. A., and Hsu, P. C. (2006). Induction of viable but non-culturable state in Vibrio cholerae O139 by temperature and its pathogenicity. *Journal of Food and Drug Analysis* 14(3), 265-272.

Sussman, M., Loya, Y., Fine, M., and Rosenberg, E. (2003). The marine fireworm Hermodice carunculata is a winter reservoir and spring-summer vector for the coral-bleaching pathogen Vibrio shiloi. *Environmental Microbiology* 5(4), 250-255.

Taylor, J. L., Tuttle, J., Pramukul, T., Obrien, K., Barrett, T. J., Jolbitado, B., Lim, Y. L., Vugia, D., Morris, J. G., Tauxe, R. V., and Dwyer, D. M. (1993). An Outbreak of Cholera in Maryland Associated with Imported Commercial Frozen Fresh Coconut Milk. *Journal of Infectious Diseases* 167(6), 1330-1335.

Telkov, M. V., Demina, G. R., Voloshin, S. A., Salina, E. G., Dudik, T. V., Stekhanova, T. N., Mukamolova, G. V., Kazaryan, K. A., Goncharenko, V., Young, M., and Kaprelyants, A. S. (2006). Proteins of the Rpf (resuscitation promoting factor) family are peptidoglycan hydrolases. *Biochemistry-Moscow* 71(4), 414-422.

Thomas, K. U., Joseph, N., Reveendran, O., and Nair, S. (2006). Salinity-induced survival strategy of Vibrio cholerae associated with copepods in Cochin backwaters. *Marine Pollution Bulletin* 52(11), 1425-1430.

Trainor, V. C., Udy, R. K., Bremer, P. J., and Cook, G. M. (1999). Survival of Streptococcus pyogenes under stress and starvation. *Fems Microbiology Letters* 176(2), 421-428.

Turner, J. W., Good, B., Cole, D., and Lipp, E. K. (2009). Plankton composition and environmental factors contribute to Vibrio seasonality. *Isme Journal* 3(9), 1082-1092.

Utsalo, S. J., Eko, F. O., Umoh, F., and Asindi, A. A. (1999). Faecal excretion of Vibrio cholerae during convalescence of cholera patients in Calabar, Nigeria. *European Journal of Epidemiology* 15(4), 379-381.

van de Giessen, A. W., Heuvelman, C. J., Abee, T., and Hazeleger, W. C. (1996). Experimental studies on the infectivity of non-culturable forms of Campylobacter spp. in chicks and mice. *Epidemiol Infect* 117(3), 463-70.

van Overbeek, L. S., Bergervoet, J. H. H., Jacobs, F. H. H., and van Elsas, J. D. (2004). The low-temperature-induced viable-but-nonculturable state affects the virulence of Ralstonia solanacearum biovar 2. *Phytopathology* 94(5), 463-469.

Varela, G., Olarte, J., Perezmir.A, and Filloy, L. (1971). Failure to Find Cholera and Noncholera Vibrios in Diarrheal Disease in Mexico-City, 1966-67. *American Journal of Tropical Medicine and Hygiene* 20(6), 925-&.

Verhoeff-Bakkenes, L., Hazeleger, W. C., de Jonge, R., and Zwietering, M. H. (2009). Campylobacter jejuni: a study on environmental conditions affecting culturability and *in vitro* adhesion/invasion. *J Appl Microbiol* 106(3), 924-31.

Verhoeff-Bakkenes, L., Hazeleger, W. C., Zwietering, M. H., and De Jonge, R. (2008). Lack of response of INT-407 cells to the presence of non-culturable Campylobacter jejuni. *Epidemiology and Infection* 136(10), 1401-1406.

Visser, I. J. R., Vellema, P., van Dokkum, H., and Shimada, T. (1999). Isolation of Vibrio cholerae from diseased farm animals and surface water in the Netherlands. *Veterinary Record* 144(16), 451-452.

von Seidlein, L., Wang, X. Y., Macuamule, A., Mondlane, C., Puri, M., Hendriksen, I., Deen, J. L., Chaignat, C. L., Clemens, J. D., Ansaruzzaman, M., Barreto, A., Songane, F. F., and Lucas, M. (2008). Is HIV infection associated with an increased risk for cholera? Findings from a case-control study in Mozambique. *Tropical Medicine & International Health* 13(5), 683-688.

Vora, G. J., Meador, C. E., Bird, M. M., Bopp, C. A., Andreadis, J. D., and Stenger, D. A. (2005). Microarray-based detection of genetic heterogeneity, antimicrobial resistance, and the viable but nonculturable state in human pathogenic Vibrio spp. *Proceedings of the National Academy of Sciences of the United States of America* 102(52), 19109-19114.

Wachsmuth, I. K., Evins, G. M., Fields, P. I., Olsvik, O., Popovic, T., Bopp, C. A., Wells, J. G., Carrillo, C., and Blake, P. A. (1993). The Molecular Epidemiology of Cholera in Latin-America. *Journal of Infectious Diseases* 167(3), 621-626.

Walker, P. J. (2005). Bovine ephemeral fever in Australia and the world. *World of Rhabdoviruses* 292, 57-80.

Watson, S. P., Clements, M. O., and Foster, S. J. (1998). Characterization of the starvation-survival response of Staphylococcus aureus. *Journal of Bacteriology* 180(7), 1750-1758.

Weichart, D., and Kjelleberg, S. (1996). Stress resistance and recovery potential of culturable and viable but nonculturable cells of Vibrio vulnificus. *Microbiology-Uk* 142, 845-853.

Weichart, D., McDougald, D., Jacobs, D., and Kjelleberg, S. (1997). In situ analysis of nucleic acids in cold-induced nonculturable Vibrio vulnificus. *Applied and Environmental Microbiology* 63(7), 2754-2758.

Weichart, D., Oliver, J. D., and Kjelleberg, S. (1992). Low temperature induced non-culturability and killing of Vibrio vulnificus. *FEMS Microbiol Lett* 79(1-3), 205-10.

Whitesides, M. D., and Oliver, J. D. (1997). Resuscitation of Vibrio vulnificus from the viable but nonculturable state. *Applied and Environmental Microbiology* 63(3), 1002-1005.

Wood, D. N., Chaussee, M. A., Chaussee, M. S., and Buttaro, B. A. (2005). Persistence of Streptococcus pyogenes in stationary-phase cultures. *Journal of Bacteriology* 187(10), 3319-3328.

Woodall, C. J. (2009). Waterborne diseases - What are the primary killers? *Desalination* 248(1-3), 616-621.

Woodward, W. E., and Mosley, W. H. (1972). Spectrum of Cholera in Rural Bangladesh .2. Comparison of El-Tor-Ogawa and Classical Inaba Infection. *American Journal of Epidemiology* 96(5), 342-&.

Worden, A. Z., Seidel, M., Smriga, S., Wick, A., Malfatti, F., Bartlett, D., and Azam, F. (2006). Trophic regulation of Vibrio cholerae in coastal marine waters. *Environmental Microbiology* 8(1), 21-29.

Wright, A. C., Hill, R. T., Johnson, J. A., Roghman, M. C., Colwell, R. R., and Morris, J. G. (1996). Distribution of Vibrio vulnificus in the Chesapeake Bay. *Applied and Environmental Microbiology* 62(2), 717-724.

Xu, H. S., Roberts, N., Singleton, F. L., Attwell, R. W., Grimes, D. J., and Colwell, R. R. (1982). Survival and Viability of Nonculturable Escherichia-Coli and Vibrio-Cholerae in the Estuarine and Marine-Environment. *Microbial Ecology* 8(4), 313-323.

Yamamoto, H. (2000). Viable but nonculturable state as a general phenomenon of non-spore-forming bacteria, and its modeling. *J Infect Chemother* 6(2), 112-4.

Yamamoto, H., Hashimoto, Y., and Ezaki, T. (1996). Study of nonculturable Legionella pneumophila cells during multiple-nutrient starvation. *Fems Microbiology Ecology* 20(3), 149-154.

Yeruham, I., Gur, Y., and Braverman, Y. (2007). Retrospective epidemiological investigation of an outbreak of bovine ephemeral fever in 1991 affecting dairy cattle herds on the Mediterranean coastal plain. *Veterinary Journal* 173(1), 190-193.

Zhong, L., Chen, J., Zhang, X. H., and Jiang, Y. A. (2009). Entry of Vibrio cincinnatiensis into viable but nonculturable state and its resuscitation. *Lett Appl Microbiol* 48(2), 247-52.

Ziprin, R. L., Droleskey, R. E., Hume, M. E., and Harvey, R. B. (2003). Failure of viable nonculturable Campylobacter jejuni to colonize the cecum of newly hatched leghorn chicks. *Avian Diseases* 47(3), 753-758.

Ziprin, R. L., and Harvey, R. B. (2004). Inability of cecal microflora to promote reversion of viable nonculturable Campylobacter jejuni. *Avian Diseases* 48(3), 647-650.

CHAPTER V

SENESCENT BACTERIA AND MODELS OF AGING

Section 1
Bacterial Senescence: A Review

Aging is inevitable in higher organisms and involves many complex biochemical processes characterized by a number of visible changes in the body. As we age, a number of changes occur to the various systems of the body causing age-related problems, many of which are irreversible. A decrease in general body function, reproductive ability and immune function is followed by many pathophysiological conditions that ultimately lead to the death of the organism. Despite the obvious disadvantages of aging, it is not opposed by natural selection probably because it can be beneficial to the species by avoiding overcrowding and promoting further evolution, thus increasing the fitness of subsequent generations (Kirkwood and Austad 2000). Different theories have been proposed to explain the complex processes of aging, but none of them can be considered complete and each theory has its own pros and cons.

Theories of aging

Biological theories of aging are broadly divided into stochastic and non-stochastic theories of aging (Kirkwood 2005; Yin and Chen 2005; Weinert and Timiras 2003; Chen et al. 2007; Hipkiss 2006; Friguet 2006). The former views aging as a phenomenon resulting from random events leading to cellular damage whereas the latter considers aging a programmed or a predetermined phenomenon occurring in all organisms within a particular time-frame. These theories can also be divided into different levels: the whole animal level (wear and tear theory, evolution theory, metabolic rate theory), the organ level (neuroendocrine theory, immunological theory), the cellular level (cell membrane theory, somatic mutation theory, mitochondrial theory, limited cellular proliferation theory) and the molecular level (free radical theory, cross linkage theory, DNA alteration theory, gene mutation theory, telomere shortening theory) (Weinert and Timiras 2003; Yin and Chen 2005).

The pogrammed theory of aging considers aging to be a predetermined or a genetic phenomenon rather than a phenomenon resulting from simple wear and tear or random accumulation of cellular damage (Prinzinger 2005). As per this theory, aging results from genes switching on and off at specific periods of time, from changes in hormone levels resulting in the loss of regulation by the hypothalamus and the decline of the immune system over time (Weinert and Timiras 2003). A number of genes including *daf-2, hsp-16, hsp-70, KLOTHO, sir2* and *sod1* are related to aging and longevity (Warner 2005; Yin and Chen 2005). However, the aging process is likely not dependent on a single gene, but is probably regulated by many different genes working together through separate pathways of aging (Warner 2005). One of the important pieces of evidence supporting the programmed theory is that the lifespan of an organism is almost fixed (Prinzinger 2005). Though it may be possible to improve lifespan to some extent through proper management, the overall lifespan of an organism is fairly constant. Prinzinger (2005) pointed out that if aging were genetically programmed, an internal clock should operate at the cellular level to control the aging process. Experiments have shown that cells have a biological limit to the number of times they can divide (Hayflick 1965), indicating that aging can be programmed at the cellular level. Similarly, the lifespan of an organism may be dependent on the absolute amount of energy that the mitochondria can generate, indicating a correlation between energy turnover and lifespan (Prinzinger 2005). As per this theory, animals that spend less energy have a longer lifespan. Thus, crocodiles and tortoises that are sluggish, birds kept in cages, animals in hibernation and mice on caloric restriction have longer lifespan than their counterparts (Prinzinger *et al.* 2005).

On the other hand, the stochastic theory suggests that the damages to the membranes, DNA, RNA, lipids and proteins that accumulate in a cell over a period of time are responsible for aging. Thus, aging can result from simple wear and tear occurring over time, damage of cellular constituents by free radicals, somatic DNA damage and accumulation of cross-linked proteins (Weinert and Timiras 2003; Kirkwood 2005; Yin and Chen 2005; Chen *et al.* 2007; Hipkiss 2006; Friguet 2006) . DNA damage accumulates with age, which can be attributed to an increase in the production of free radicals and to a reduced DNA repair capacity (Chen *et al.* 2007; Lu and Finkel 2008; Lu *et al.* 2004). Increased DNA damages can result in premature senescence (Chen *et al.* 2004; Chen *et al.* 2005; Chen *et al.* 2007). However, there are also reports indicating that oxidative damage of DNA may not always correlate with the lifespan of an organism (Van Remmen *et al.* 2003; Keaney *et al.* 2004; Miwa *et al.* 2004; Vermulst *et al.* 2007). Another factor responsible for aging is the accumulation of altered proteins, which is considered a hallmark of cellular senescence (Hipkiss 2006; Friguet 2006; Dukan *et al.* 2000; Maisonneuve *et al.* 2008a). Proteins can be altered or modified by oxygen and reactive nitrogen species, metabolic aldehydes and lipid peroxidation products (Hipkiss 2006). Even though altered proteins also occur normally in cells, they are selectively degraded by proteases. The age-related accumulation of these proteins can be either due to their increased production or due to decreased protein repair and removal (Hipkiss 2006; Friguet 2006). Among the different modifications, protein carbonylation induced by oxygen free-radicals, lipid peroxidation and reducing sugars is the most common (Dukan *et al.* 2000; Maisonneuve *et al.* 2008a, b). Carbonylated proteins may gradually form aggregates that ultimately become less degradable (Maisonneuve *et al.* 2008b). Even healthy cells may contain some carbonylated proteins that are then passed on to the next generation. However, over time, the non-degradable aggregates of carbonylated proteins accumulate and eventually lead to cell death (Maisonneuve *et al.* 2008b).

Since none of these theories is sufficient to explain all the mechanisms of the aging phenomenon, a 'network theory' of aging has been proposed that integrates the contributions of various mechanisms in the different theories together (Kirkwood *et al.* 2003; Kirkwood 2005). For example, DNA damages and mutations that have accumulated over time can affect mitochondrial function, which in turn may produce large amounts of free radicals that destroy cellular constituents and accelerate aging (Kirkwood *et al.* 2003; Kirkwood 2005). Similarly, the metabolic theory can also be linked to the mitochondrial or free radical theory since the absolute energy produced by mitochondria may determine the lifespan of an organism.

Role of mitochondria

Free oxygen radicals are normally produced in cells as a result of oxidative phosphorylation. The mitochondrion is the organelle in which oxidative phosphorylation occurs and is also the main source of free oxygen radicals and H_2O_2. Moreover, mitochondria regulate stress response and apoptosis (Salvioli *et al.* 2001). A substantial body of evidence supports a role for oxidative damage to the mitochondrial respiratory chain and mitochondrial DNA (mtDNA) in the determination of mammalian lifespan (Trifunovic *et al.* 2004; Martin and Loeb 2004; Lee and Wei 2007). About 70-80% of mitochondrial O_2^- is released into the mitochondrial matrix which, if not removed, can lead to oxidative stress and damage (Navarro and Boveris 2007). Intra-mitochondrial superoxide dismutase (SOD) catalyzes the reaction that helps to remove O_2^- and form H_2O_2, which is then converted to H_2O with the help of the enzyme glutathione peroxidase. Overexpression of MnSOD in adult *Drosophila melanogaster* reduced the superoxide level and increased the average lifespan of flies by 16% (Sun *et al.* 2002). On the other hand, ablation of mitochondrial SOD2 in *D. melanogaster* increased the endogenous oxidative stress that resulted in accelerated aging (Kirby *et al.* 2002).

Aging may reduce mitochondrial function by: reducing mitochondrial membrane potential, the expression of the electron transport pathway and H^+-driven ATP synthesis; increasing H^+ permeability, mitochondrial production of reactive oxygen species, accumulation of deletion and point mutations of mtDNA; and by inducing apoptosis (Salvioli *et al.* 2001; Balaban *et al.* 2005; Lee and Wei 2007; Trifunovic *et al.* 2004; Martin and Loeb 2004; Kujoth *et al.* 2005; Navarro and Boveris 2007). The transcription profiles of aging in humans, mice and flies have shown that aging in all three organisms is associated with the decreased expression of electron transport pathways, indicating a common aging signature (Zahn *et al.* 2006). Oxidative damage is negatively correlated with the activities of mitochondrial complexes I and IV, and aging reduces the activities of these complexes, which in turn leads to an increased rate of O_2^- generation (Navarro and Boveris 2007). Thus, oxidative damage creates a vicious cycle wherein the initial ROS mediated damage leads to increased oxidant production which causes further damage to mitochondria (Balaban *et al.* 2005; Linnane *et al.* 1989).

The involvement of mitochondria in aging is further highlighted in a study wherein mice exhibited accelerated aging and had a shortened lifespan when their mitochondrial DNA was made error prone by eliminating its proofreading ability (Trifunovic *et al.* 2004; Martin and Loeb 2004; Prinzinger 2005). Without the proofreading activity of DNA polymerase, errors in mitochondrial DNA replication remain uncorrected, leading to the accumulation of deletion and point mutations and resulting in accelerated aging (Trifunovic *et al.* 2004). Similarly, mice expressing error-prone mitochondrial DNA polymerase g accumulated mtDNA mutations and exhibited accelerated aging, which was attributed to increased apoptosis and subsequent loss of irreplaceable cells (Kujoth *et al.* 2005). Skeletal muscles in aged humans showed reduced mtDNA, mRNA amounts and mitochondrial ATP production, whereas the level of the oxidative DNA lesion 8-oxo-deoxyguanosine was increased, supporting the oxidative theory (Short *et al.* 2005; Shigenaga *et al.* 1989). Thus, various experimental results suggest mitochondria as the seat of senescence; therefore, the lifespan of an organism may depend on mitochondrial function.

Bacterial growth and division

Growth and multiplication of bacteria is the basis of turbidity of liquid media and colony formation on solid media. It is also the basis of bacterial infections in which bacteria multiply and produce toxins that affect the various organs and the body as a whole. Bacteria such as *E. coli* divide once in 20 min, if adequate nutrients are present in the medium. There are typically four phases or stages of bacterial growth. Initially, their number remains low and during this lag phase, bacteria adapt to the new conditions and environment. This is followed by the log phase where they multiply exponentially utilizing the nutrients and then by a stationary phase where their number remains almost constant. The stationary phase results from the depletion of nutrients, change in pH or the accumulation of waste materials. During this phase, the number of bacteria newly formed is in equilibrium with the number of bacteria undergoing death (Kolter *et al.* 1993). The final stage is the death phase, where more bacteria begin to die, resulting in a gradual decline in their numbers.

Bacteria divide by binary fission, which also occurs in different stages. Plasmid and chromosomal DNA replication and segregation are followed by cytokinesis, during which a septum is formed at the middle, dividing the bacteria into two (Slater and Schaechter 1974). Once DNA replication has been completed, cell division or cytokinesis begins with the development of a septum at the midpoint of a pre-divisional cell (Slater and Schaechter 1974). It proceeds with the invagination of the cytoplasmic membrane, the peptidoglycan wall and cell wall material at the central plane of the cell accompanied by the synthesis of septal peptidoglycans (Slater and Schaechter 1974; Lutkenhaus and Addinall 1997) and ultimately pinches off giving rise to two progeny cells. At the cell center where the future division occurs, a contractile ring, called the Z-ring, is formed (Harry 2001; Lutkenhaus and Addinall 1997). In bacteria, cell division involves the positioning of the Z-ring whereas in fungal and animal cells, it involves the positioning of the contractile ring (Lutkenhaus and Addinall 1997). In *E. coli*, at least ten different proteins (which are cytoplasmic, periplasmic and membrane proteins) are required, which localize to the constriction site (Vicente *et al.* 2006; Buddelmeijer and Beckwith 2002; Rothfield *et al.* 1999). The assembly of proteins into this ring is initiated by the FtsZ protein, which is structurally and functionally analogous to tubulin protein and is the most abundant of the cell division proteins (Rothfield *et al.* 1999). It has GTP-ase activity and polymerizes *in vitro* in a GTP-dependent manner (de Boer *et al.* 1992; Vicente *et al.* 2006). FtsZ functions as the universal prokaryotic division protein and is also widespread in eubacteria, archeae, and some eukaryotic organelles (Vicente *et al.* 2006; Bramhill 1997). However, it is absent in mitochondria, where FtsZ has been replaced with dynamin (Erickson 2000; Vicente *et al.* 2006). Mutation in *ftsZ* results in the failure of septum formation leading to the generation of filaments (Addinall and Lutkenhaus 1996), whereas its overexpression induces minicell formation by increasing the septation events per unit cell mass (Ward and Lutkenhaus 1985). During the early stages of the cell division cycle, FtsZ is present throughout the cytoplasm (Vicente *et al.* 2006). But as the cell elongates, it localizes to the Z-ring. Once FtsZ is localized to the Z-ring, other cell division proteins like FtsA, FtsI, FtsK, FtsL, ZipA etc., are recruited in a sequential and almost linear fashion (Vicente *et al.* 2006). FtsZ, not only helps in the assembly of divisional apparatus but also transduces cell-cycle specific signals that catalyze the septal invagination (Rothfield *et al.* 1999). Initiation of cell division may not occur until all the proteins are localized (Buddelmeijer and Beckwith 2002). Once all the proteins are assembled, Z-ring contraction occurs, resulting in cell division. Once cell division is over, the septal ring dissembles as a morphological unit (Vicente *et al.* 2006; Bramhill 1997).

The divisional site at the midcell region in *E. coli* is determined by a set of proteins of the *minCED* gene locus (de Boer *et al.* 1989; Rothfield *et al.* 1999; Vicente *et al.* 2006) and by nucleoid occlusion, mediated by Noc in *B. subtilis* and slmA in *E. coli* (Vicente *et al.* 2006). In the absence of *min* gene products, septation fails to occur at the midcell but rather occurs at regions adjacent to the cell pole, resulting in the formation of minicells (Rothfield *et al.* 1999). Nucleoid exclusion prevents the assembly of proteins at regions where a nucleoid is present (Vicente *et al.* 2006). In *Noc* mutants, division occurs through the nucleoid, whereas in wild type, division occurs in regions devoid of nucleoid (Wu and Errington 2004). Thus, the function of *Noc* is to direct FtsZ assembly away from the nucleoid and to protect the nucleoid from damage by means of the division septum (Wu and Errington 2004).

Bacterial senescence

For more than a century, bacteria were considered to be functionally immortal organisms because of their symmetrical division. A parent bacterium split symmetrically into two equal daughter cells, and both the cells receive damaged and new cellular constituents. Thus, bacteria were considered to be organisms immune to natural aging and death, even though they can be killed by many external agents like starvation or other stressors. But recent findings prove that they undergo death not only due to starvation or exposure to stressors but also by replicative senescence (Stewart *et al.* 2005; Ackermann *et al.* 2003).

As a bacterial culture moves from the exponential growing phase to the stationary phase, cells are exposed to low nutrient and toxic environments. This may induce changes in bacterial metabolic activities and protein expression patterns (Kolter *et al.* 1993; Nystrom 2004). They gradually accumulate oxidative damage and divide slowly. However, on transfer to fresh nutrient medium, they may revert to normal growth. On the other hand, if a bacterial culture is exposed to the stationary phase for prolonged periods, their ability to recover from the damages may be lost and the cells may finally undergo death. This conditional senescence (Nystrom 2003; Fredriksson and Nystrom 2006) is different from the replicative senescence in that the latter results from the sequential loss of fitness following multiple rounds of replication (Fredriksson and Nystrom 2006). Replicative senescence, therefore, suggests that a bacterium cannot be functionally immortal and that it also undergoes normal aging and death even if all essential nutrients are available for its growth.

Conditional senescence

E. coli grown in culture medium enters the stationary phase once the exogenous nutrients become limited. Their reproductive ability, manifested as the ability to form colonies on solid medium, gradually reduces and is finally lost completely. The average lifespan of a stationary phase *E. coli* is around 3-5 days when exogenous carbon becomes unavailable to the cells (Ericsson *et al.* 2000; Matin 1991). Even though the death is not due to replicative senescence, the conditional senescence (Nystrom 2002, 2003, 2004; Fredriksson and Nystrom 2006) exhibits many similarities with the aging process of higher organisms, including oxidation of cellular components, redistribution of resources from reproductive functions to cellular maintenance, and stress resistance and repair (Nystrom 2004). The life expectancy of stationary-phase *E. coli* cells is correlated with the level of oxidative damage incurred, similar to aging in higher organisms (Nystrom 2004).

One of the hallmarks of conditional senescence is the decrease in the size of the bacteria resulting from reductive division and dwarfing (Kolter et al. 1993; Nystrom 2004). Reductive division occurs during the stationary phase of bacterial growth. When exogenous energy sources are limited, as in the stationary phase, the organism continues to reproduce using the energy from endogenous sources while simultaneously using previously initiated multiple replication forks to continue DNA synthesis (Nystrom 2004). This results in an increase in cell number without an increase in the cellular biomass (Kolter et al. 1993; Baker et al. 1983; Hood et al. 1986). Thus, most of the cells produced during reductive division will be smaller and coccoid in appearance (Baker et al. 1983; Hood et al. 1986; Watson et al. 1998) and this change in cell morphology can be attributed to the differential expression of certain genes (Santos et al. 2002). Stationary phase bacteria increasingly express the bolA morphogene, which in turn affects the morphology of cells through the transcriptional control of other genes like dacA, dacC, and ampC (Santos et al. 2002). Dwarfing, on the other hand, is a continuous reduction in cell size after completion of reductive division and is induced by starvation (Nystrom 2004). Dwarfing may allow the cell a maximum chance of survival during long-term starvation (Humphrey et al. 1983; Kjelleberg et al. 1983). It is a form of self-digestion that results in the degradation of cellular components like the cell envelope or the cell wall (Nystrom 2004); however, the DNA content may remain almost the same (Bakken and Olsen 1989).

Another bacterial response to starvation is reduction in the bacterial growth and metabolic rate (Kolter et al. 1993). Even though the overall rate of protein synthesis decreases in the stationary phase, a distinct set of proteins is induced, many of which function to protect the cell from external stressors (Kolter et al. 1993). Hence the stationary phase cells or starved cells are resistant to a variety of other stresses as well (Jouper-Jaan et al. 1992; Matin 1991). This cross-protection is largely dependent on two regulatory networks, σ^S, the general stress defense regulon, and σ^{32}, the heat shock regulon (Kolter et al. 1993; Lange and Hengge-Aronis 1991; Fredriksson and Nystrom 2006). A number of proteins expressed during starvation, including bolA, depend on σ^S, whose activity in turn is regulated by a variety of starvation conditions (Kolter et al. 1993). On the other hand, σ^{32} induces the expression of heat shock proteins like DnaK and GroEL (Kolter *et al.* 1993; Fredriksson and Nystrom 2006). Mutants lacking either σ^S or σ^{32} exhibit accelerated senescence with markedly elevated levels of protein carbonylation, a protein modification suggestive of oxidative damage (Kolter *et al.* 1993; Dukan and Nystrom 1998). The stationary-phase induced growth arrest in bacteria results in increased protein oxidation including protein carbonylation and oxidative disulfide bridge formation (Dukan and Nystrom 1998; Nystrom 2002). Specific proteins like EF-Tu, EF-G, Hsp-70 chaperone, pyruvate kinase, DnaK and several TCA cycle enzymes are more susceptible to protein oxidation (Dukan and Nystrom 1998; Nystrom 2002). Conditional senescence in the stationary phase is correlated with the level of protein oxidation since the over-expression of SOD could mitigate the induction of protein oxidation and increase the lifespan of an organism (Dukan and Nystrom 1998).

Replicative senescence

In the 1960's, Leonard Hayflick found that normal diploid cells have a limited lifespan (Hayflick 1965). They lose the ability to divide after a certain number of cell divisions. This phenomenon is called replicative senescence, the "Hayflick phenomenon", or the "Hayflick limit" and reflects the aging process of the whole organism. Replicative senescence has been studied in model eukaryotic organisms like yeast and *Drosophila* (Kennedy *et al.* 1994; Kern *et al.* 2001). Recently, aging has been described in prokaryotes like bacteria as well (Stewart *et al.* 2005).

Simple eukaryotes such as the yeast *Saccharomyces cerevisiae* divides asymmetrically wherein the daughter cell that buds from the mother cell is smaller in size (Hartwell and Unger 1977). The mother cell divides a finite number of times before it stops dividing completely, indicating that the yeast cells undergo aging and have a specific lifespan (Mortimer and Johnston 1959). Lifespan in yeast can be measured in terms of replicative or chronological lifespan (MacLean *et al.* 2001). Replicative lifespan is measured as the number of daughter cells produced by a mother cell before it stops dividing, whereas the chronological lifespan is measured as the length of time a yeast cell remains viable in the stationary phase (MacLean *et al.* 2001). Kennedy *et al.* (1994) studied the replicative aging of the budding yeast *S. cerevisiae* by microscopically following the mother cells through a number of cell divisions. They noticed that the mother cell underwent a finite number of divisions and that the size of both the mother cell and the daughter cell increased with age. They also noticed that older mother cells underwent a more symmetrical division in which the daughter cell that budded from the mother cell was almost the same size as the mother cell at division. Similarly, the daughter cells of older mother cells had a shorter lifespan than those of younger mother cells. The daughter cells of older mother cells in the last 10% of their lifespan was found to undergo only 7.9 divisions, whereas those daughter cells budded from the mother cells in the first 70% of the lifespan, on an average, divided 26.5 times. Symmetrical division, therefore, does not give any advantage to the daughter cells as they undergo a smaller number of divisions than the mother cell. In the case of the fruit fly D. *melanogaster* also, hatching success rate and offspring viability decrease with parental age (Kern *et al.* 2001). This decline in offspring viability was apparent only in older females and appeared to be a general feature of *Drosophila* senescence.

Until recently, replicative senescence has been described in eukaryotes only. Perhaps, the earliest evidence of replicative senescence in bacteria was provided by Liu (1999). By tracking the bacterial growth in liquid media with high viscosity, Liu (1999) observed the unidirectional growth and reproduction of E. *coli*. He proposed that the bacterium has an intrinsic cell polarity with one end behaving as a mother compartment and the other end as the daughter compartment resulting in the formation of two bacteria of succeeding generations. His model defined bacterial age by its experienced chronological time. Based on this model, he predicted that, on bacterial division, the old strand of DNA remain with the mother bacterium whereas the new strand goes to the daughter bacterium and that this distribution of old and new strands of DNA between the mother and daughter cells is responsible for the intrinsic differences between the two. However, this finding was not accepted or appreciated readily at the time.

Replicative senescence was later reported in *Caulobacter crescentus*, a bacterium in which cytokinesis is intrinsically asymmetrical (Ackerman *et al.* 2003). The bacterium can freely swim in water and this free swimming cell represents the swarmer cell. The swarmer cell is non-reproductive, but after a period of free swimming, it gets differentiated to a sessile reproductive stalked cell. The stalked cell remains attached to the substrate but produces progeny swarmer cells which then separates from the stalked cells and begin free swimming. Ackerman *et al.* (2003) studied the replicative senescence in *Caulobacter crescentus* by using microscopy flow chambers in which the stalked cells were attached to the chamber while the swarmer cells produced from the stalked cells were removed by the medium flowing through the chamber. They noticed that, as the stalked cells gave rise to more progenies, their rate of division slowed and finally stopped completely. However, the progenies produced, i.e. the swarmer cells, were rejuvenated offspring. Moreover, those progenies produced towards the end were indistinguishable from any of the young cells produced earlier in the experiment in terms of cell division time. These experiments proved that the stalked cells undergo senescence and that their reproductive output decreases with increasing aging. However, the major difference was that, unlike *S. cerevisiae* and *D. melanogaster*, the progenies from both the older and younger mother cells were rejuvenated offspring and indistinguishable from each other.

In contrast to *C. crescentus*, where cytokinesis is intrinsically asymmetrical, most bacteria including *E. coli* reproduces by morphologically symmetrical division. Stewart *et al.* (2005) studied the senescence in *E. coli* using automated time lapse microscopy by following repeated cycles of reproduction. They followed individual exponentially growing cells up to nine generations of growth and reproduction. Their findings were comparable to those reported previously by Liu (1999). The bacterium exhibits cell polarities which give rise to an old pole with a reduced growth rate and a new pole with higher growth rate. They found that the average growth rate of old pole cells was 2.2% slower than that of new pole cells and that the new pole cells were larger and divided sooner than the old pole cells. In addition, the old pole cells were also more likely to die than the new pole cells. They concluded that the two apparently identical cells are functionally asymmetrical, with the old pole cell behaving as the aging mother cell and the new pole cell as the rejuvenated offspring. As in the case of *C. crescentus*, the bacterial lineage can be maintained through the young rejuvenated offspring.

Thus, a common feature of both prokaryotic and eukaryotic senescence is that the older mother cells, after a number of divisions, gradually lose their reproductive ability and finally stop division and undergo death. However, a major difference is that, in prokaryotes (*E. coli.* and *C. crescentus*), the daughter cells formed from both younger and older mother cells are rejuvenated offspring (Ackermann *et al.* 2003; Stewart *et al.* 2005; Liu 1999) whereas in eukaryotes (*D. melanogaster* and *S. cerevisiae*), the daughter cells from the older mother cells have lower lifespan than those from the younger mother cells (Kennedy *et al.* 1994; Kern *et al.* 2001; Hartwell and Unger 1977).

Asymmetrical cellular damage

Another common feature of aging is the asymmetrical segregation of damaged macromolecules in the mother cell. Aging is associated with the accumulation of damaged proteins (Desnues *et al.* 2003, Aguilaniu *et al.* 2003; Erjavec *et al.* 2007), genetic materials (Sinclair and Guarente 1997) or dysfunctional mitochondria (Lai *et al.* 2002).

Cells of an *E. coli* population show asymmetry not only with respect to growth rate, but also with respect to protein oxidation levels (Desnues *et al.* 2003; Aguilaniu *et al.* 2003; Erjavec *et al.* 2007). An *E. coli* population consists of relatively low damaged daughter cells (low protein oxidation) that are reproductively competent and damaged mother cells with reduced reproductive ability (Desnues *et al.* 2003). In *S. cerevisiae* as well, the levels of protein oxidation increase with the replicative age of mother cells (Aguilaniu *et al.* 2003; Erjavec *et al.* 2007). Oxidized proteins are unevenly distributed between the mother and daughter cells during cytokinesis, with mother cells retaining most of the oxidized proteins (Aguilaniu *et al.* 2003). The ability of mother cells to retain oxidized proteins is dependent on Sir2p (NAD+-dependent histone deacetylase, a regulator of aging in many organisms), since the oxidized proteins are distributed equally between the mother and daughter cells in Sir2 mutants (Aguilaniu *et al.* 2003). With increasing age, mother cells lose their ability to retain the oxidized proteins (Aguilaniu *et al.* 2003). The oxidized proteins accelerate aging by forming aggregates that can cause cytotoxicity (Maisonneuve *et al.* 2008a). In exponentially growing *E. coli*, the amount of protein aggregates increases over time and were found to be more prevalent in dead cells than in culturable cells (Maisonneuve *et al.* 2008a). Similarly, aggregated proteins accumulate in cells with older poles, which are associated with a decrease in reproductive ability (Lindner *et al.* 2008).

In yeast cells, extrachromosomal rDNA circles (ERC) are associated with the aging process (reviewed by Guarente 1996; Guarente 1997). The yeast ribosomal DNA consists of many tandemly arrayed copies of a 9.1-kb repeat, and the homologous recombination between rDNA repeats results in ERCs (reviewed by Sinclair and Guarente 1997). ERCs can self-replicate but usually segregate to mother cells as they lack centromeric DNA sequences. Because they are segregated to mother cells, the amount of ERCs increases with the aging of mother cells and finally results in nucleolar enlargement, fragmentation and replicative senescence and death (Sinclair and Guarente 1997). Even though they are not inherited by daughter cells, once a threshold level of ERCs accumulates in the mother cell, asymmetry may break down and even the daughter cell may start to inherit some of them (Sinclair and Guarente 1997).

Similarly, segregation of dysfunctional mitochondria may be a feature of aging in *S. cerevisiae* (Lai *et al.* 2002). Mutation of ATP2, the gene encoding the beta-subunit of mitochondrial F_oF_1-ATPase, abrogated the age asymmetry between the mother and the daughter cells (Lai *et al.*2002). In normal wild type yeast cells, mother cells retain most of the dysfunctional mitochondria themselves, but in mutants, the lack of segregation of active mitochondria to daughter cells results in the accumulation of dysfunctional mitochondria in daughter cells, leading to their accelerated aging (Lai *et al.* 2002). Mitochondrial oxidative damage, resulting from the accumulation of free oxygen radicals, is one of the main signs of aging. SOD, which removes the free oxygen radicals, can improve the lifespan of many organisms (Sun *et al.* 2002; Kirby *et al.* 2002). Harris *et al.* (2003) noticed that the overexpression of MnSOD increased the chronological lifespan but decreased the replicative lifespan of *S. cerevisiae*. The increase in the chronological lifespan could be due to the reduced oxidative stress, whereas the decrease in chronological age could be due to mitochondrial defects in older mother cells which may occur independently of Sir2p, indicating that chronological aging and replicative aging may occur through independent pathways (Harris *et al.* 2003).

Asymmetrical distribution of damaged macromolecules may have evolutionary advantages. By retaining the damaged macromolecules itself, the mother cell spares the daughter cell from accumulating damage. This would help maintain the lineage at the expense of the mother cell (Ackermann *et al.* 2007).

Conclusion

E. coli, which reproduces by morphologically symmetrical division, also undergoes replicative senescence and death. Mother cells gradually lose their reproductive ability and finally stop division and undergo death. However, the daughter cells formed from mother cells are rejuvenated offspring. During cytokinesis, damaged or oxidized proteins are unevenly distributed between the mother and daughter cells, with mother cells retaining most of the oxidized proteins, thus sparing the daughter cells from accumulating damages.

Section 2
Why the Current Model of *E. Coli* Aging is Incomplete

The relationship between persisters and senescent bacteria
When an antibiotic-sensitive bacterial culture is treated with penicillin, the majority of the population may die, but a small subpopulation may somehow survive. This small subpopulation of bacteria is called persisters (Lewis 2007). Persisters have already been reviewed before; however, some of the features of persisters are given below.

Persisters neither grow in the presence of antibiotics nor are killed by antibiotics. However, on removal of antibiotics, they grow and repopulate the culture. This population is again sensitive to antibiotics except for a small subpopulation. They are not mutants nor do they carry any antibiotic-resistant genes. They exhibit a phenotypic trait which attributes tolerance to antibiotics. But this tolerance is transient and is not passed on to their progeny. The number of persisters is low during the exponential phase of growth, but increases as the culture enters the stationary phase. They also exhibit cross-tolerance to other antibiotics. The biphasic killing pattern exhibited by a bacterial culture is due to the presence of persisters.

Integrating the features of persistence and bacterial senescence, Klapper *et al.* (2007) proposed that persisters are senescent bacteria. Their proposal is based on the model of bacterial senescence put forward by Stewart *et al.* (2005). As per the model, mother cells undergo gradual aging and have a reduced growth rate and finally stop dividing, whereas the daughter cell produced from a mother cell is a rejuvenated offspring capable of faster growth. Klapper *et al.* (2007) also made an assumption that the older cells are more tolerant to antibiotics than the younger cells due to their slow growth rate.

Thus, as per their proposal, when a bacterial culture is treated with bactericidal antibiotics, the younger cells are killed due to their fast growth rate whereas the older mother cells (persisters) survive. However, upon removal of antibiotics, the rejuvenated offspring produced from the mother cells quickly repopulate the culture. Since, during the exponential phase, the number of older cells is low, there may not be many survivors when antibiotics are used against exponential phase bacteria. Thus, Kappler *et al.* (2005) argued that senescence can explain all the features of persister cells. Their argument would have been correct if the current model of bacterial aging were true. However, the current model of bacterial aging may not be complete.

Why is the current model incomplete?
One drawback with the current model of aging is that it cannot explain WHY the presence of mother cells in a bacterial culture is advantageous to the whole population. From an evolutionary standpoint, the aging process is always considered disadvantageous to any species even though it is naturally selected (Kirkwood and Austad 2000). In higher organisms like humans, reproductive ability is completely lost after a certain age. Similarly, advanced maternal age is associated with decreased reproductive ability, increased miscarriages, poor oocyte quality and chromosomal abnormalities (Munne *et al.* 1995; Armstrong 2001). In the case of simple eukaryotes also, the daughter cells from the older mother cells exhibit a slower growth rate and accelerated aging (Kennedy *et al.* 1994; Kern *et al.* 2001; Hartwell and Unger 1977). Thus, there is a natural restriction on reproduction in older mother cells in order to avoid the consequences of 'abnormal offspring'. However, the *E. coli* senescence model put forward by Liu (1999) and Stewart *et al.* (2005) proposes that the older mother cells give rise to rejuvenated offspring. If their model is correct, senescence is disadvantageous only to the individual mother cell but is advantageous to the population because old mother cells not only can give rise to rejuvenated offspring but also are resistant to antibiotics. The main reason for this discrepancy can be attributed to the inability to track *E. coli* growth and division for many generations. Stewart *et al.* (2005) could track *E. coli* division for only nine generations, and the results from the nine generations were extrapolated to the whole life period of *E. coli*. Their model may apply only for the first few generations, and thus the current model may not be complete. I propose that *E. coli* also follows the same aging model as that of yeast cells; i.e., as mother cells undergo aging, they exhibit a slower growth rate and accumulate damaged macromolecules. Younger mother cells segregate damaged macromolecules themselves and give rise to new, rejuvenated daughter cells. However, after a number of divisions, mother cells lose their control over segregation and thereafter the daughter cells also inherit damaged

molecules, resulting in their accelerated aging. From this point onwards, the division becomes almost symmetrical. Thus, if we start with a single old mother cell at the terminal stages of senescence, the whole lineage will undergo senescence.

I also propose that persisters are senescent bacteria; however, my definition for persisters and the model of *E. coli* aging are different. This will be discussed more in the following chapters.

Conclusion

The aging model of *E. coli* is comparable to that of simple eukaryotes. Towards the terminal stages of division, the mother cell loses their control over segregation of damaged macromolecules and thereafter the daughter cells also inherit damaged molecules, resulting in their accelerated aging. Thus, the division becomes symmetrical in the case of mother cells at the terminal stages of senescence.

References

Ackermann, M., Chao, L., Bergstrom, C. T., and Doebeli, M. (2007). On the evolutionary origin of aging. *Aging Cell* 6(2), 235-44.

Ackermann, M., Stearns, S. C., and Jenal, U. (2003). Senescence in a bacterium with asymmetric division. *Science* 300(5627), 1920.

Addinall, S. G., Bi, E., and Lutkenhaus, J. (1996). FtsZ ring formation in fts mutants. *J Bacteriol* 178(13), 3877-84.

Aguilaniu, H., Gustafsson, L., Rigoulet, M., and Nystrom, T. (2003). Asymmetric inheritance of oxidatively damaged proteins during cytokinesis. *Science* 299(5613), 1751-3.

Armstrong, D. T. (2001). Effects of maternal age on oocyte developmental competence. *Theriogenology* 55(6), 1303-22.

Baker, R. M., Singleton, F. L., and Hood, M. A. (1983). Effects of nutrient deprivation on Vibrio cholerae. *Appl Environ Microbiol* 46(4), 930-40.

Bakken, L. R., and Olsen, R. A. (1989). DNA-Content of Soil Bacteria of Different Cell-Size. *Soil Biology & Biochemistry* 21(6), 789-793.

Balaban, R. S., Nemoto, S., and Finkel, T. (2005). Mitochondria, oxidants, and aging. *Cell* 120(4), 483-95.

Bramhill, D. (1997). Bacterial cell division. *Annu Rev Cell Dev Biol* 13, 395-424.

Buddelmeijer, N., and Beckwith, J. (2002). Assembly of cell division proteins at the E. coli cell center. *Curr Opin Microbiol* 5(6), 553-7.

Chen, J. H., Hales, C. N., and Ozanne, S. E. (2007). DNA damage, cellular senescence and organismal ageing: causal or correlative? *Nucleic Acids Res* 35(22), 7417-28.

Chen, J. H., Ozanne, S. E., and Hales, C. N. (2005). Heterogeneity in premature senescence by oxidative stress correlates with differential DNA damage during the cell cycle. *DNA Repair (Amst)* 4(10), 1140-8.

Chen, J. H., Stoeber, K., Kingsbury, S., Ozanne, S. E., Williams, G. H., and Hales, C. N. (2004). Loss of proliferative capacity and induction of senescence in oxidatively stressed human fibroblasts. *J Biol Chem* 279(47), 49439-46.

de Boer, P., Crossley, R., and Rothfield, L. (1992). The essential bacterial cell-division protein FtsZ is a GTPase. *Nature* 359(6392), 254-6.

de Boer, P. A., Crossley, R. E., and Rothfield, L. I. (1989). A division inhibitor and a topological specificity factor coded for by the minicell locus determine proper placement of the division septum in E. coli. *Cell* 56(4), 641-9.

Desnues, B., Cuny, C., Gregori, G., Dukan, S., Aguilaniu, H., and Nystrom, T. (2003). Differential oxidative damage and expression of stress defence regulons in culturable and non-culturable Escherichia coli cells. *EMBO Rep* 4(4), 400-4.

Dukan, S., Farewell, A., Ballesteros, M., Taddei, F., Radman, M., and Nystrom, T. (2000). Protein oxidation in response to increased transcriptional or translational errors. *Proc Natl Acad Sci U S A* 97(11), 5746-9.

Dukan, S., and Nystrom, T. (1998). Bacterial senescence: stasis results in increased and differential oxidation of cytoplasmic proteins leading to developmental induction of the heat shock regulon. *Genes Dev* 12(21), 3431-41.

Erickson, H. P. (2000). Dynamin and FtsZ. Missing links in mitochondrial and bacterial division. *J Cell Biol* 148(6), 1103-5.

Ericsson, M., Hanstorp, D., Hagberg, P., Enger, J., and Nystrom, T. (2000). Sorting out bacterial viability with optical tweezers. *J Bacteriol* 182(19), 5551-5.

Erjavec, N., Larsson, L., Grantham, J., and Nystrom, T. (2007). Accelerated aging and failure to segregate damaged proteins in Sir2 mutants can be suppressed by overproducing the protein aggregation-remodeling factor Hsp104p. *Genes Dev* 21(19), 2410-21.

Fredriksson, A., and Nystrom, T. (2006). Conditional and replicative senescence in Escherichia coli. *Curr Opin Microbiol* 9(6), 612-8.

Friguet, B. (2006). Oxidized protein degradation and repair in ageing and oxidative stress. *FEBS Lett* 580(12), 2910-6.

Guarente, L. (1996). Do changes in chromosomes cause aging? *Cell* 86(1), 9-12.

Guarente, L. (1997). Link between aging and the nucleolus. *Genes Dev* 11(19), 2449-55.

Harris, N., Costa, V., MacLean, M., Mollapour, M., Moradas-Ferreira, P., and Piper, P. W. (2003). Mnsod overexpression extends the yeast chronological (G(0)) life span but acts independently of Sir2p histone deacetylase to shorten the replicative life span of dividing cells. *Free Radic Biol Med* 34(12), 1599-606.

Harry, E. J. (2001). Bacterial cell division: regulating Z-ring formation. *Mol Microbiol* 40(4), 795-803.

Hartwell, L. H., and Unger, M. W. (1977). Unequal division in Saccharomyces cerevisiae and its implications for the control of cell division. *J Cell Biol* 75(2 Pt 1), 422-35.

Hayflick, L. (1965). The Limited *in vitro* Lifetime of Human Diploid Cell Strains. *Exp Cell Res* 37, 614-36.

Hipkiss, A. R. (2006). On the mechanisms of ageing suppression by dietary restriction-is persistent glycolysis the problem? *Mech Ageing Dev* 127(1), 8-15.

Hood, M. A., Guckert, J. B., White, D. C., and Deck, F. (1986). Effect of nutrient deprivation on lipid, carbohydrate, DNA, RNA, and protein levels in Vibrio cholerae. *Appl Environ Microbiol* 52(4), 788-93.

Humphrey, B., Kjelleberg, S., and Marshall, K. C. (1983). Responses of Marine Bacteria Under Starvation Conditions at a Solid-Water Interface. *Appl Environ Microbiol* 45(1), 43-47.

Jouperjaan, A., Goodman, A. E., and Kjelleberg, S. (1992). Bacteria Starved for Prolonged Periods Develop Increased Protection against Lethal Temperatures. *Fems Microbiology Ecology* 101(4), 229-236.

Keaney, M., Matthijssens, F., Sharpe, M., Vanfleteren, J., and Gems, D. (2004). Superoxide dismutase mimetics elevate superoxide dismutase activity *in vivo* but do not retard aging in the nematode Caenorhabditis elegans. *Free Radic Biol Med* 37(2), 239-50.

Kennedy, B. K., Austriaco, N. R., Jr., and Guarente, L. (1994). Daughter cells of Saccharomyces cerevisiae from old mothers display a reduced life span. *J Cell Biol* 127(6 Pt 2), 1985-93.

Kern, S., Ackermann, M., Stearns, S. C., and Kawecki, T. J. (2001). Decline in offspring viability as a manifestation of aging in Drosophila melianogaster. *Evolution* 55(9), 1822-31.

Kirby, K., Hu, J., Hilliker, A. J., and Phillips, J. P. (2002). RNA interference-mediated silencing of Sod2 in Drosophila leads to early adult-onset mortality and elevated endogenous oxidative stress. *Proc Natl Acad Sci U S A* 99(25), 16162-7.

Kirkwood, T. B. (2005). Understanding the odd science of aging. *Cell* 120(4), 437-47.

Kirkwood, T. B., and Austad, S. N. (2000). Why do we age? *Nature* 408(6809), 233-8.

Kirkwood, T. B., Boys, R. J., Gillespie, C. S., Proctor, C. J., Shanley, D. P., and Wilkinson, D. J. (2003). Towards an e-biology of ageing: integrating theory and data. *Nat Rev Mol Cell Biol* 4(3), 243-9.

Kjelleberg, S., Humphrey, B. A., and Marshall, K. C. (1983). Initial Phases of Starvation and Activity of Bacteria at Surfaces. *Appl Environ Microbiol* 46(5), 978-984.

Klapper, I., Gilbert, P., Ayati, B. P., Dockery, J., and Stewart, P. S. (2007). Senescence can explain microbial persistence. *Microbiology* 153(Pt 11), 3623-30.

Kolter, R., Siegele, D. A., and Tormo, A. (1993). The stationary phase of the bacterial life cycle. *Annu Rev Microbiol* 47, 855-74.

Kujoth, G. C., Hiona, A., Pugh, T. D., Someya, S., Panzer, K., Wohlgemuth, S. E., Hofer, T., Seo, A. Y., Sullivan, R., Jobling, W. A., Morrow, J. D., Van Remmen, H., Sedivy, J. M., Yamasoba, T., Tanokura, M., Weindruch, R., Leeuwenburgh, C., and Prolla, T. A. (2005). Mitochondrial DNA mutations, oxidative stress, and apoptosis in mammalian aging. *Science* 309(5733), 481-4.

Lai, C. Y., Jaruga, E., Borghouts, C., and Jazwinski, S. M. (2002). A mutation in the ATP2 gene abrogates the age asymmetry between mother and daughter cells of the yeast Saccharomyces cerevisiae. *Genetics* 162(1), 73-87.

Lange, R., and Hengge-Aronis, R. (1994). The cellular concentration of the sigma S subunit of RNA polymerase in Escherichia coli is controlled at the levels of transcription, translation, and protein stability. *Genes Dev* 8(13), 1600-12.

Lee, H. C., and Wei, Y. H. (2007). Oxidative stress, mitochondrial DNA mutation, and apoptosis in aging. *Exp Biol Med (Maywood)* 232(5), 592-606.

Lewis, K. (2007). Persister cells, dormancy and infectious disease. *Nat Rev Microbiol* 5(1), 48-56.

Lindner, A. B., Madden, R., Demarez, A., Stewart, E. J., and Taddei, F. (2008). Asymmetric segregation of protein aggregates is associated with cellular aging and rejuvenation. *Proc Natl Acad Sci U S A* 105(8), 3076-81.

Linnane, A. W., Marzuki, S., Ozawa, T., and Tanaka, M. (1989). Mitochondrial DNA mutations as an important contributor to ageing and degenerative diseases. *Lancet* 1(8639), 642-5.

Liu, S. V. (1999). Tracking bacterial growth in liquid media and a new bacterial life model. *Science in China* 42, 644-654.

Lu, T., and Finkel, T. (2008). Free radicals and senescence. *Exp Cell Res* 314(9), 1918-22.

Lu, T., Pan, Y., Kao, S. Y., Li, C., Kohane, I., Chan, J., and Yankner, B. A. (2004). Gene regulation and DNA damage in the ageing human brain. *Nature* 429(6994), 883-91.

Lutkenhaus, J., and Addinall, S. G. (1997). Bacterial cell division and the Z ring. *Annu Rev Biochem* 66, 93-116.

MacLean, M., Harris, N., and Piper, P. W. (2001). Chronological lifespan of stationary phase yeast cells; a model for investigating the factors that might influence the ageing of postmitotic tissues in higher organisms. *Yeast* 18(6), 499-509.

Maisonneuve, E., Ezraty, B., and Dukan, S. (2008). Protein aggregates: an aging factor involved in cell death. *J Bacteriol* 190(18), 6070-5.

Maisonneuve, E., Fraysse, L., Lignon, S., Capron, L., and Dukan, S. (2008). Carbonylated proteins are detectable only in a degradation-resistant aggregate state in Escherichia coli. *J Bacteriol* 190(20), 6609-14.

Martin, G. M., and Loeb, L. A. (2004). Ageing: mice and mitochondria. *Nature* 429(6990), 357-9.

Matin, A. (1991). The molecular basis of carbon-starvation-induced general resistance in Escherichia coli. *Mol Microbiol* 5(1), 3-10.

Miwa, S., Riyahi, K., Partridge, L., and Brand, M. D. (2004). Lack of correlation between mitochondrial reactive oxygen species production and life span in Drosophila. *Ann N Y Acad Sci* 1019, 388-91.

Mortimer, R. K., and Johnston, J. R. (1959). Life span of individual yeast cells. *Nature* 183(4677), 1751-2.

Munne, S., Alikani, M., Tomkin, G., Grifo, J., and Cohen, J. (1995). Embryo morphology, developmental rates, and maternal age are correlated with chromosome abnormalities. *Fertil Steril* 64(2), 382-91.

Navarro, A., and Boveris, A. (2007). The mitochondrial energy transduction system and the aging process. *Am J Physiol Cell Physiol* 292(2), C670-86.

Nystrom, T. (2002). Aging in bacteria. *Curr Opin Microbiol* 5(6), 596-601.

Nystrom, T. (2003). Conditional senescence in bacteria: death of the immortals. *Mol Microbiol* 48(1), 17-23.

Nystrom, T. (2004). Stationary-phase physiology. *Annu Rev Microbiol* 58, 161-81.

Prinzinger, R. (2005). Programmed ageing: the theory of maximal metabolic scope. How does the biological clock tick? *EMBO Rep* 6 Spec No, S14-9.

Rothfield, L., Justice, S., and Garcia-Lara, J. (1999). Bacterial cell division. *Annu Rev Genet* 33, 423-48.

Salvioli, S., Bonafe, M., Capri, M., Monti, D., and Franceschi, C. (2001). Mitochondria, aging and longevity--a new perspective. *FEBS Lett* 492(1-2), 9-13.

Santos, J. M., Lobo, M., Matos, A. P., De Pedro, M. A., and Arraiano, C. M. (2002). The gene bolA regulates dacA (PBP5), dacC (PBP6) and ampC (AmpC), promoting normal morphology in Escherichia coli. *Mol Microbiol* 45(6), 1729-40.

Shigenaga, M. K., Gimeno, C. J., and Ames, B. N. (1989). Urinary 8-hydroxy-2'-deoxyguanosine as a biological marker of *in vivo* oxidative DNA damage. *Proc Natl Acad Sci U S A* 86(24), 9697-701.

Short, K. R., Bigelow, M. L., Kahl, J., Singh, R., Coenen-Schimke, J., Raghavakaimal, S., and Nair, K. S. (2005). Decline in skeletal muscle mitochondrial function with aging in humans. *Proc Natl Acad Sci U S A* 102(15), 5618-23.

Sinclair, D. A., and Guarente, L. (1997). Extrachromosomal rDNA circles--a cause of aging in yeast. *Cell* 91(7), 1033-42.

Slater, M., and Schaechter, M. (1974). Control of cell division in bacteria. *Bacteriol Rev* 38(2), 199-221.

Stewart, E. J., Madden, R., Paul, G., and Taddei, F. (2005). Aging and death in an organism that reproduces by morphologically symmetric division. *PLoS Biol* 3(2), e45.

Sun, J., Folk, D., Bradley, T. J., and Tower, J. (2002). Induced overexpression of mitochondrial Mn-superoxide dismutase extends the life span of adult Drosophila melanogaster. *Genetics* 161(2), 661-72.

Trifunovic, A., Wredenberg, A., Falkenberg, M., Spelbrink, J. N., Rovio, A. T., Bruder, C. E., Bohlooly, Y. M., Gidlof, S., Oldfors, A., Wibom, R., Tornell, J., Jacobs, H. T., and Larsson, N. G. (2004). Premature ageing in mice expressing defective mitochondrial DNA polymerase. *Nature* 429(6990), 417-23.

Van Remmen, H., Ikeno, Y., Hamilton, M., Pahlavani, M., Wolf, N., Thorpe, S. R., Alderson, N. L., Baynes, J. W., Epstein, C. J., Huang, T. T., Nelson, J., Strong, R., and Richardson, A. (2003). Life-long reduction in MnSOD activity results in increased DNA damage and higher incidence of cancer but does not accelerate aging. *Physiol Genomics* 16(1), 29-37.

Vermulst, M., Bielas, J. H., Kujoth, G. C., Ladiges, W. C., Rabinovitch, P. S., Prolla, T. A., and Loeb, L. A. (2007). Mitochondrial point mutations do not limit the natural lifespan of mice. *Nat Genet* 39(4), 540-3.

Vicente, M., Rico, A. I., Martinez-Arteaga, R., and Mingorance, J. (2006). Septum enlightenment: assembly of bacterial division proteins. *J Bacteriol* 188(1), 19-27.

Ward, J. E., Jr., and Lutkenhaus, J. (1985). Overproduction of FtsZ induces minicell formation in E. coli. *Cell* 42(3), 941-9.

Warner, H. R. (2005). Developing a research agenda in biogerontology: basic mechanisms. *Sci Aging Knowledge Environ* 2005(44), pe33.

Watson, S. P., Clements, M. O., and Foster, S. J. (1998). Characterization of the starvation-survival response of Staphylococcus aureus. *J Bacteriol* 180(7), 1750-8.

Weinert, B. T., and Timiras, P. S. (2003). Invited review: Theories of aging. *J Appl Physiol* 95(4), 1706-16.

Wu, L. J., and Errington, J. (2004). Coordination of cell division and chromosome segregation by a nucleoid occlusion protein in Bacillus subtilis. *Cell* 117(7), 915-25.

Yin, D., and Chen, K. (2005). The essential mechanisms of aging: Irreparable damage accumulation of biochemical side-reactions. *Exp Gerontol* 40(6), 455-65.

Zahn, J. M., Sonu, R., Vogel, H., Crane, E., Mazan-Mamczarz, K., Rabkin, R., Davis, R. W., Becker, K. G., Owen, A. B., and Kim, S. K. (2006). Transcriptional profiling of aging in human muscle reveals a common aging signature. *PLoS Genet* 2(7), e115.

CHAPTER VI

INTEGRATING PERSISTERS, SCV, VBNC, AND SENESCENT BACTERIA WITH PK/PD PARAMETERS

Section 1
Persisters Show Heritable Phenotypes and Generate Bacterial Heterogeneity and Noise in Protein Expression

This non-peer-reviewed article is posted in Nature Precedings and is available at http://precedings.nature.com/documents/1411/version/2. Below is the same after some proofreading.

Abstract

Persisters are a small subpopulation of bacteria that survive lethal concentrations of an antibiotic without having antibiotic resistance genes. The isolation of persisters from a normally dividing population is considered difficult due to their slow growth, low numbers and phenotypic shift; i.e., when re-grown in antibiotic-free medium, they revert to their parent population. The inability to isolate persisters is a major hindrance in this field of research. The 'phenotypic shift' of persisters observed previously is questioned here. Persisters, on the other hand, may exhibit a heritable phenotype and hence can be easily isolated from a normally dividing population by allowing their selective growth. Rather than a single subset, they comprise many distinct subgroups, each exhibiting different growth rates, colony sizes, antibiotic tolerance and protein expression levels. Clearly, they are one of the sources of bacterial heterogeneity and noise in protein expression. Existence of persisters in a normally dividing population can explain some of the unsolved puzzles like antibiotic tolerance, the post-antibiotic effect and the viable but non-culturable bacterial state. It is hypothesized that persisters are aging bacteria.

Key words: Persisters, bacterial aging, post-antibiotic effect, small colony variants, antibiotic tolerance, viable but non-culturable bacteria

Persisters were first described by Joseph Bigger (1944) when he found that a culture of *Staphylococcus* spp. was not completely sterilized by a lethal concentration of ampicillin. Most of the bacteria were lysed by ampicillin, but a small subpopulation somehow survived. When the surviving bacteria were grown in an antibiotic free medium, they grew just like the parent population and were again susceptible to ampicillin. These surviving bacteria or persisters either grew slowly or did not grow at all in the presence of antibiotics but reverted to normal growth upon removal of antibiotics (Balaban *et al.* 2004; Lewis 2007). Since the persisters reverted to their original population upon removal of antibiotics, they were considered to be dormant bacteria that avoided killing by antibacterial agents (Lewis 2007). This phenotypic switch is widely considered to be a survival strategy of bacteria against antibiotics (Lewis 2007). Due to this phenotypic shift, it is difficult to isolate persisters from a normally dividing population (Lewis 2007). Hence most of the studies on persisters are done with *E. coli* mutant hipA, which is reported to produce a high frequency of persisters (Black *et al.* 1991; Korch *et al.* 2006; Moyed and Bertrand 1983; Moyed and Broderick 1986).

Single persister cells were studied using microfluidic devices which identified at least two types of persisters (Balaban *et al.* 2004). Type-I persisters are formed in the stationary phase and thus constitute a pre-existing population of non-growing cells. However, they revert to growing cells in an antibiotic-free medium with an extended time lag. On the other hand, Type-II persisters do not originate from the stationary phase but are continuously generated during the growth phase. Thus a wild type population consists of normal cells, in addition to Type-I and Type-II persisters (Balaban *et al.* 2004).

Persisters are not mutants nor induced by the antibiotics, but are preformed in a culture (Balaban *et al.* 2004; Lewis 2007). During the lag and early exponential phase of growth, persister formation is very low, but their number increases during the mid to late exponential phase of growth (Keren *et al.* 2004). By growing bacteria repeatedly at early exponential phases of growth, persisters can be eliminated completely (Keren *et al.* 2004).

Persisters may be responsible for the resistance of biofilms to antibiotics (Brooun *et al.* 2000; Spoering and Lewis 2001). A biofilm is formed when bacteria grow on a surface and is enclosed in an exopolymer matrix. Biofilms are notorious for recalcitrant infections and resistance to antibiotics (Al-Fattani and Douglas 2006; Costerton *et al.* 1999) and are responsible for more than 65% of infections in the West (Costerton *et al.* 2003). Isolation of persisters from a normally dividing population may help in understanding the pathogenesis of biofilm resistance to antibiotics.

Here it is reported that bacterial populations consist of persisters that are tolerant to antibiotics. However, they show a heritable phenotype, rather than exhibiting a phenotypic shift. The 'phenotypic shift' of persisters reported earlier can be due to improper experimental settings. Persisters can be isolated from a normally dividing population by utilizing the 'concentration dependent killing' property of aminoglycosides. Persisters thus isolated show many significant differences from those reported earlier apart from the lack of phenotypic shift. They comprise many subgroups that are formed during all phases of bacterial growth. Each subgroup exhibits its own characteristic properties, including a difference in protein expression levels. The existence of persisters is suggested to be one of the important sources of bacterial heterogeneity and noise in protein expression. They can be aging bacteria and hence can be isolated from most bacterial cultures.

Materials and methods

Bacterial strains, growth conditions, antibiotics and chemicals
The bacterial strains used were *E. coli* DH5α and BL-21(DE-3) cells (Invitrogen). *Salmonella enterica* serovar Typhimurium was a generous gift from Dr. Suman Mukhopadhyay, VA-MD Regional College of Veterinary Medicine. *E. coli* DH5α with pET 14-B plasmid carrying GFP was obtained from Dr. Iqbal Hamza, University of Maryland, College Park. All antibiotics (kanamycin, ampicillin, tetracycline and nalidixic acid) were obtained from Sigma. Concentrations of stock solution of the antibiotics were kanamycin-10mg/ml, ampicillin-50mg/ml, tetracycline-5 mg/ml and nalidixic acid-15 mg/ml.

Luria-Bertani (LB) broth and LB agar base were used to culture the above cells. The chemicals used were glycine, Tris (Fischer), acrylamide, TEMED, SDS, Ammonium per sulphate, bromophenol blue (Biorad), BCA protein assay reagents (PIERCE), thiamine, thymidine (sigma), and mercaptoethanol (Invitrogen).

Isolation of persisters using kanamycin
For isolating pure cultures of persisters, 50 μl of overnight culture of *E. coli* DH5α was added to 3 ml of fresh LB medium containing kanamycin at concentrations of 10, 20, 30, 40 and 50 μg/ml and incubated at 37°C for 48-60 h at 240 r.p.m. When the culture reached an O.D. of approximately 0.5, 200 μl was withdrawn and centrifuged to remove the supernatant containing antibiotics. The pellet was washed with fresh LB medium twice. After the second wash, approximately equal concentrations of bacteria were added to 3 ml of LB medium containing the same concentration of kanamycin and incu-

bated as before. The whole procedure was repeated once again to ensure that persister culture was not contaminated by any normal cells. The growth rate of bacteria was determined from the O.D. measured at definite time periods using a nanodrop ND-1000 spectrophotometer. The measurable range of O.D. by nanodrop spectrophotometer is 0.01-0.2, which is almost equivalent to the O.D. value of 0.1-2 in the classical cuvette-based systems, as per the manufacturer's manual. Hence, the O.D. value obtained in the nanodrop spectrophotometer will be multiplied by a factor of 10 to make the reading equivalent to the cuvette-based system.

For growth on agar, 50 µl of stationary phase culture was serially diluted 1 in 10 times up to 10^8 concentration. 30 µl from each dilution was plated on LB agar and incubated at 37°C for 48-60 hrs. Plates containing uniformly spread colonies were selected. The same method was followed for other antibiotics. Colonies on agar were then photographed using the chemidoc system.

For the heritability test, 50 µl of the isolated persisters were centrifuged to remove the supernatant containing antibiotics, and the pellet was washed with fresh LB medium without antibiotics twice. Approximately equal concentrations of bacteria, determined from O.D. of the culture, from each group of persisters were added to 3ml of fresh LB medium without antibiotics and incubated. The growth rate was determined as above. They were plated onto agar as described previously. The whole procedure was repeated twice to test the heritability.

Persister differences between antibiotics

50 µl of overnight culture of E. coli DH5α was diluted to a concentration of 10^{-4} to 10^{-5} and plated on LB agar. A hole was punched at the centre of the agar using a sterile pipette. Kanamycin (10 mg/ml) was added to fill the hole and the plate was then incubated at 37°C for 48 h. The same method was followed for ampicillin (50 mg/ml), tetracycline (5mg/ml) and nalidixic acid (10 mg/ml).

Effect of initial inoculum size and time of exposure of antibiotic

50, 100, 250, 500 and 1,000 µl of overnight culture were added to fresh LB medium to make a final volume of 3 ml. 50 µg of kanamycin was added to each and incubated as before. After 48 h, approximately equal concentrations of bacteria (determined from O.D. values) were withdrawn, pelleted and washed with fresh LB twice and added to a final volume of 3 ml LB medium and incubated as before. Growth rate was determined as above.

To study the effect of time of exposure of antibiotics, 50 μl of overnight culture was added to 3 ml LB medium containing 50 μg/ml of kanamycin and incubated at 37°C as before. At the designated time periods, samples containing approximately equal concentrations of bacteria (determined by O.D. values) were withdrawn and added to fresh LB medium without kanamycin to a final volume of 3 ml and incubated as before. The growth rate was determined as before.

Persister formation and induction by antibiotics
To determine whether persisters are induced by antibiotics, 50 μl of overnight stationary culture was added to 3 ml of LB medium and incubated as above. After reaching the early exponential phase (O.D. approximately 0.5), 100 μl of the culture was withdrawn and added to fresh medium, then incubated again at the same conditions. This cycle was repeated 4 times. 100 μl of the mid-exponential phase culture from each cycle was added to 3 ml of fresh medium containing kanamycin at different concentrations as before and incubated for 48 h and then the growth rate was measured. The mid exponential culture from the fourth cycle was allowed to reach a stationary phase by incubating overnight. This overnight culture with approximately equal concentrations of bacteria as above was added to 3 ml of fresh medium containing the same concentrations of kanamycin as above and incubated as before.

Auxotrophy of persisters
50 μl of isolated persisters (K-30 and K-40) were added to LB medium containing thiamine (Sigma) and thymidine (Sigma) at concentrations of 1, 3, 5, 10 and 20 μg/ml by adding these chemicals to LB medium and incubated as above. The growth rate was determined as before. They were also plated to agar containing thiamine and thymidine at these concentrations to test for reversion to large colonies.

Tolerance of persisters to different antibiotics
Approximately equal concentrations of bacterial cultures of normal and persister cells were treated with the same and higher concentrations of kanamycin at which they were isolated. Persisters were also

treated with different concentrations of ampicillin and nalidixic acid. They were incubated for 48 h as before.

Expression of GFP determined by SDS polyacrylamide gel electrophoresis and western blotting

pET-14B vector carrying GFP in *E. coli* DH5α was transformed to BL21 (DE-3) cells by the heat shock method. Persisters of BL21 (DE-3) cells were then isolated using kanamycin as before. Persisters thus isolated were grown in antibiotic-free medium for 48 h as described before. When persisters grown in antibiotic-free medium reached the stationary phase, they were pelletted, resuspended in PBS and heated at 95°C for 10 minutes. A small volume was used to determine the total protein concentration by BCA protein assay (Pierce) with bovine serum albumin as the standard and the rest pelletted and stored at -20°C. For immunoblotting, stored pellets were heated at 95°C for 2 minutes in the presence of 1xSDS loading buffer containing DTT. Samples were run on 12% SDS–PAGE and transferred onto a nitrocellulose membrane (Biorad) using semi-dry apparatus (Biorad) at 9V for 45 minutes. Blots were blocked overnight using 5% non-fat dry milk and incubated with an antibody against green fluorescent protein (Invitrogen) at a concentration of 1:5,000 for 1.5 h followed by incubation with a secondary antibody (goat anti-rabbit conjugated with horseradish peroxidase (biorad)). The signal was detected using the Amersham western blotting detection kit (GE healthcare) and the blot was quantified by the chemidoc system. Protein bands were detected by Coomassie staining and the molecular weight of the proteins was estimated using marker proteins (Kaleidoscope standards-Biorad).

Expression of GFP by immunofluorescence

The intensity of GFP fluorescence in the normal wild type and persisters of BL21 (DE-3) cells were visualized from the fluorescent images captured on a Leica DMIRE2 microscope fitted with a Q-imaging camera using Simple PCI software.

Statistical analysis

The data presented are given in mean ± standard error of mean (s.e.m.). The *n* for each data set is given below each table.

Results and discussion

Isolation of persisters

For the isolation of persisters, the 'concentration dependent killing' property of kanamycin (Craig and Ebert 1990; Freeman *et al.* 1997; Tam *et al.* 2006), an aminoglycoside that kills bacteria by inhibiting protein synthesis, is utilized here. Bacteria are killed by kanamycin in a growth rate dependent manner. At a low concentration, it kills fast-dividing bacteria only, leaving the persisters, which are slow-growers. At high concentrations, slow-growing cells also become susceptible, but again in a concentration-dependent manner. This property of the antibiotic is utilized here to separate bacteria with different growth rates. The different groups of persisters will be designated as K-10, K-20, K-30 etc., depending on the concentration of kanamycin in μg/ml used for isolation of persisters. Among the persisters, K-10 had the highest growth rate and K-50 the least (Table.1). At a kanamycin concentration of 60 μg/ml, no growth could be detected. Since persisters divide slowly, they take more time to reach turbidity, which is evident from the optical density (O.D.) readings of the spectrophotometer. Extreme persisters like K-40 or K-50 never reach turbidity. They reached a stationary phase before growing to turbidity. The above pattern of growth was observed with both *E. coli* DH5-α and BL21 (DE-3) cells. With *Salmonella enteritica* serovar Typhimurium LT2, the growth pattern was similar except that there was no growth of bacteria at a kanamycin concentration above 30 μg/ml (Table. 1).

When plated on agar, they exhibited properties consistent with growth in liquid medium (Fig.1.A, B and C). The time it took for colonies to appear on agar increased with higher kanamycin concentrations. K-40 required 40-48 h to form visible colonies on agar. There was a clear difference in the size of the colonies, which decreased proportionally with higher concentrations of the antibiotic. Again, consistent with growth in liquid medium, the size of colonies of K-30 and K-40 did not increase much even after many days of incubation.

Table 1. Growth characteristics of persisters isolated by kanamycin

Bacterial strain	KAN* (ug/ml)	O.D. of bacterial culture						Colony Growth (h)$^\Delta$
		In presence of antibiotics			Upon removal of antibiotics§			
		9h	24h	48h	9h	24h	48h	
E.coli DH-5α	0	t	t	t	t	t	t	>15
	10	0.23±0.02	t	t	1.04±0.09	t	t	>15
	20	>0.1	0.82±0.08	t	0.43±0.04	t	t	17±1.2
	30	>0.1	0.14±0.02	0.53±0.05	0.16±0.02	0.74±0.05	0.82±0.06	24±1.3
	40	>0.1	>0.1	0.18±0.02	>0.1	0.1±0.02	0.25±0.03	48±1.7
	50	>0.1	>0.1	0.16±0.01	>0.1	0.1±0.01	0.2±0.02	53±5.6
	60	>0.1	>0.1	>0.1	>0.1	>0.1	>0.1	-
S.enteritica	0	t	t	t	t	t	t	>15
	10	>0.1	t	t	t	t	t	>15
	20	>0.1	0.36±0.03	0.77±0.07	0.21±0.04	1.03±0.08	1.31±0.14	25±2.4
	30	>0.1	>0.1	0.12±0.01	>0.1	>0.1	0.15±0.01	32±3.6
	40	>0.1	>0.1	>0.1	>0.1	>0.1	>0.1	-

Results are shown as mean ± s.e.m.; $n=3$ for *E.coli* DH-5α; $n=2$ for *S.enteritica.*

t – turbid

*Concentration of kanamycin (ug/ml) used for isolation of persisters.

§ persisters isolated were washed twice and incubated in fresh LB medium for 48 h. This procedure was repeated.

$^\Delta$Time taken to form visible colonies after antibiotic-free persister cultures were plated on agar.

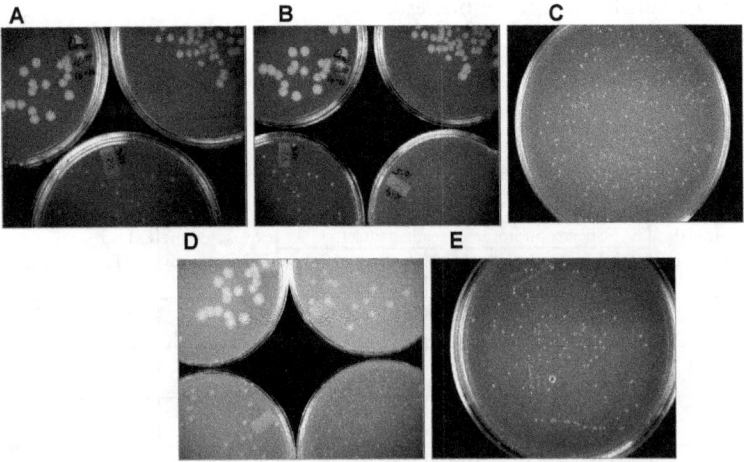

Figure 1. Persisters comprise many subpopulations having different growth rates. A. Colonies of DH5-α normal cells and persisters isolated by kanamycin after 18 h following agar plating. No colonies were formed by K-30 and K-40 at 18 h. Upper left - control, upper right - K-10 and bottom - K-20. B. colonies of control, K-10, K-20 and K-30 after 24 h. Upper left - control, upper right - K-10, lower left - K-20 and lower right - K-30. C. Colonies of K-40 after 50 h. D. Colonies of control, K-10, K-20 and K-30 grown in antibiotic-free medium after 26 h following agar plating. Upper left - control, upper right - K-10, lower left - K-20 and lower right - K-30. E. colonies of K-40 grown in antibiotic-free medium after 2 days. F. colonies of K-40 after incubation for 21 days followed by plating on agar.

Two factors are important in the isolation of persisters without 'contamination' by normal fast-dividing, wild type bacteria: 1. low initial inoculum size; and 2. prolonged time of exposure to antibiotics (Table. 2). As the initial inoculum size increases, the chance for bacteria to escape killing increases. The presence of a few normal cells that escape killing can completely change the picture as they have a definite growth advantage over persisters; hence it is important that all susceptible bacteria be killed, leaving only the persisters. This can be ensured by repeatedly treating persisters at exponential phases of growth with the same concentration of kanamycin used to isolate them. 50 μl of stationary phase bacterial culture was used for isolation of persisters, which was the ideal inoculum size for *E. coli* and *S. enteritica*. Similarly, an extended time period is also necessary to avoid 'contamination' by normal cells and to allow sufficient time for the growth of persisters. A short exposure time below 9 h did not kill all susceptible bacteria.

Table 2. Effect of inoculum size and exposure time on persister isolation

Initial inoculum size					Total time of exposure to antibiotics		
Initial inoculum size(µl)§	Spectrophotometer readings (O.D.)				exposure time to kanamycin (50 µg/ml)	O.D. after removal of kanamycin	
	With kanamycin (50 µg/ml)		After removal of kanamycin				
	24 h	48 h	24 h	48 h		24 h	48 h
50	>0.1	0.18±0.01	>0.1	0.25±0.02	1.5 h	t	t
100	>0.1	0.22±0.01	>0.1	0.23±0.02	3 h	t	t
250	0.74±0.12(L)	t(L)	t	t	6 h	t	t
500	t(L)	t(L)	t	t	9 h	t	t
1000	t(L)	t(L)	t	t	24 h	>0.1	0.22±0.02
					48 h	>0.1	0.24±0.01

Results are shown as mean ± s.e.m.; $n = 3$

t – turbid; L – lysis; t(L) – turbid with lysis

§ overnight culture of *E. coli* DH-5α

It is important to know whether the growth characteristics exhibited by different persisters are induced by the antibiotic itself. Earlier, it was reported that persisters are not induced by antibiotics (Keren *et al*. 2004). However it was also assumed that persisters comprise a single subpopulation which starts to appear in the early exponential phase, is followed by a sharp increase in the mid-exponential phase, and reaches a maximum in the stationary phase, forming approximately 1% of the total population. The experiment was repeated since a number of subgroups of persisters were isolated rather than a single subset. For this purpose, the procedure by Keren *et al*. (2004) was followed with some modifications as described in the methods section. By repeatedly growing bacteria at the early-exponential phase of the cycle, bacteria with high growth rates were selected while slow growing ones were gradually diluted out. As the number of cycles increased, there was a gradual reduction in the number of persisters (Table 3). By the fourth cycle, K-30, K-40 and K-50 were almost completely eliminated, but K-10 and K-20 still remained. However, all persisters reappeared in cultures from the fourth cycle incubated overnight. This experiment demonstrates that persisters are normally present in a bacterial population and they are not induced by antibiotics. Similarly, they comprise a number of subgroups and are formed during all phases of bacterial growth. In fact, the stage of bacterial growth does not influence the formation of persisters. It depends only on the rate of growth of individual persister groups. Though all persister groups are present in a normal population, all of them cannot be detected in agar unless they are selected by eliminating the fast-dividing bacteria. They have a definite growth disadvantage in the presence of normal wild type bacteria due to their slow growth and low numbers. Thus, extreme persisters like K-40 or K-50 can be included under viable but non-culturable bacteria (VBNC), bacteria that remain viable but cannot be cultured under normal circumstances (Bogosian and Bourneuf 2001; Nystrom 2001; Oliver 2005).

Table 3. Elimination of persisters and their re-growth[§]

Kanamycin (μg/ml)	control	cycle 1	cycle 2	cycle 3	cycle 4	stationary phase of cycle 4
0	t	t	t	t	t	t
10	t	t	t	t	t	t
20	t	1.8 ± 0.3	1.4 ± 0.2	1.18 ± 0.2	0.83 ± 0.07	t
30	0.66 ± 0.06	0.54 ± 0.08	0.33 ± 0.07	0.1 ± 0.02	>0.1	1.1 ± 0.27
40	0.22 ± 0.03	0.22 ± 0.03	0.11 ± 0.03	>0.1	>0.1	0.44 ± 0.08
50	0.12 ± 0.01	0.13 ± 0.01	>0.1	>0.1	>0.1	0.19 ± 0.04
60	>0.1	>0.1	>0.1	>0.1	>0.1	0.12 ± 0.05

Results are shown as mean $\pm$ s.e.m.; $n = 3$

t – turbid

[§] An early exponential phase culture of *E. coli* was incubated, diluted in fresh medium and reinoculated for four cycles. Bacterial culture from each cycle was treated with kanamycin at varying concentrations and incubated for 48 h. The exponential phase culture from the fourth cycle was allowed to reach the stationary phase by incubating overnight and was then treated with the same concentrations of kanamycin.

Heritability of persisters

As the persisters showed visible differences in their growth patterns in both liquid and solid medium, the heritability of the persisters was tested by growing them in antibiotic-free medium. Persisters grown in LB medium without antibiotics showed the same pattern of growth rate as the parent persister population, even though the growth of persisters was faster when compared to their growth in the presence of the antibiotic (Table 1). This faster growth rate is due to the higher initial inoculum size of the persisters. In solid medium, there were differences in the total time taken to form visible colonies, colony sizes and the growth rate of colonies (Table 1 and Fig. 1.D and E). Similarly, when the largest colony from each group was picked and grown in antibiotic-free medium for 48 h followed by plating on agar, the same pattern was again observed. This was true for at least three generations. A small colony was never seen to revert back to a larger one.

Phenotypic shift by bacteria

Since persisters were reported to exhibit a phenotypic shift, the phenomenon needs to be studied more precisely. For this purpose, bacteria were grown in a medium containing lethal concentrations of antibiotics as well as other stressors like low pH and high pH. It was observed that at lethal concentrations of any of these stressors, most of the bacteria were killed and there was no visible growth (Table 4). However, when the bacteria were transferred to medium without the stressors, they showed luxuriant growth just like normal cells. These bacteria were not mutants as they were again susceptible to the same concentration of stressors previously used. This experiment showed that the 'phenotypic shift' reported is not a property seen with antibiotics alone, but rather with any stressors. It can be assumed that, due to the lysis of the majority of the bacteria by lethal concentrations of antibiotics, the growth medium might be accumulating materials (probably some quorum-sensing molecules) released from those lysed bacteria which inhibit the growth of others and transiently protect them from the action of antibiotics. To test this hypothesis, bacteria were grown in the presence of lethal concentrations of ampicillin (15μg/ml). Once most of the bacteria were killed and there was no visible growth, they were pelletted, washed twice and then grown in fresh medium as well as in medium containing the same concentration of ampicillin (15μg/ml) previously used to kill them and then incubated for another round. While turbidity was noticed in ampicillin-free medium after overnight incubation, no growth was detected in the medium containing ampicillin. It was evident that all bacteria were killed by the second round of ampicillin treatment since no growth was detected after re-incubation following removal of ampicillin (data not shown). If bacteria were exhibiting a 'true phenotypic shift', they would have escaped the bactericidal action of ampicillin for the second round also, since they were growing under the same conditions of growth as the first round. This experiment may also explain why a greater number of bacteria survive when a high initial inoculum size is used. With a high inoculum size, the amount of materials released by lysed bacteria is also higher which may inhibit the growth of a greater number of bacteria, resulting in more survivors.

Table 4. The effect of pH on growth characteritics of *E. coli* persisters

Conditions of growth§	pH	O.D. of bacterial culture after 6 h of incubation	
		In the presence of stressors	Upon removal of stressors
Optimal pH	7.2	t	t
Acidic pH	5.1	>0.1	t
	4.6	>0.1	t
	3.5	>0.1	>0.1
Alkaline pH	8.2	>0.1	t
	8.8	>0.1	t
	10.6	>0.1	>0.1

Results are shown as mean ± s.e.m.; $n = 2$
§Overnight culture of 50 µl was used for incubation

Most of the experiments demonstrating phenotypic shift were done after incubating bacteria for a short period of time (Keren *et al.* 2004). For ampicillin and other antibiotics that exhibit a 'time-dependent killing property', it is highly important to incubate bacteria for a longer time. It is well documented that for such antibiotics, the time of incubation is more critical than the concentration of antibiotics (Craig and Ebert 1990; Jacobs 2001; Craig 1998). This is very important since a short incubation cannot kill all susceptible bacteria and hence re-inoculation may show a 'false phenotypic shift'. Even for kanamycin, a 'concentration-dependent killing' antibiotic, not all susceptible bacteria are killed by incubating for a short period (Table 2). Moreover, earlier experiments with persisters used a high initial inoculum size by growing bacteria to their exponential phase (Keren *et al.* 2004). As noted earlier, a high inoculum size could also give false results since the survival of some normal cells is sufficient to give a false phenotypic shift (Table 2). Similarly, a very high concentration of antibiotics above MIC can result in a 'paradoxical effect' wherein a significant number of bacteria escapes killing by antibiotics (Woolfrey *et al.* 1987; Woolfrey and Enright 1990) due to mechanisms that are not clear. Previously it was demonstrated that maximal killing of bacteria occurs over a narrow range of antibiotic concentrations above MIC and that the number of survivors increases with higher concentrations above this value (Woolfrey *et al.* 1987; Woolfrey and Enright 1990). Since some of the experiments with persisters might have used high concentrations of ampicillin above MIC (Balaban *et al.* 2004; Keren *et al.* 2004), the chance of giving a 'false phenotypic shift' is higher.

The mechanism of persister formation is largely unknown. It was shown that cells over-expressing HipA, RelE and other toxin-antitoxin (TA) modules resulted in the formation of a high frequency of persisters and hence these proteins were implicated in persister formation (Lewis 2007). These proteins may increase the persister population by slowing down or stopping cell division and thus evade the action of antibiotics (Lewis 2007). However, Vazquez-Laslop *et al.* (2006) questioned the specific roles of HipA and RelE in the formation of persisters. They found that proteins such as DnaJ and PmrC that are unrelated to TA modules can also result in a high frequency of persisters when they are over-expressed in cells. They concluded that when cells are expressing proteins to toxic levels, the frequency of persisters increases regardless of the kind of proteins.

Based on the above findings, it is assumed that the phenotypic shift of persisters observed earlier may be due to faulty experimental settings.

Persister formation by other antibiotics

The procedure used to isolate persisters with kanamycin was repeated with tetracycline, ampicillin and nalidixic acid. A growth pattern similar to kanamycin could not be seen with any of these antibiotics (Table 5). With a high concentration of tetracycline, O.D. did not change much after a period of time, indicating that there was no bacterial growth. With lower concentrations, the medium showed turbidity. When bacteria were plated on agar, they grew like normal cells and the size of the colonies was similar to that of the control (data not shown). This was not surprising as tetracycline is bacteriostatic and hence bacteria that remained after tetracycline treatment could revert to a normal population once the antibiotic was removed. With ampicillin and nalidixic acid, a lot of bacterial lysis was seen in liquid medium within 3 h of incubation. At low antibiotic concentrations, the surviving bacteria resumed growth and reached turbidity. However, at high concentrations, there were no signs of growth. When surviving cells were incubated in fresh medium without antibiotics, the effect was an all or none phenomenon; i.e., either they grew to turbidity or they did not grow at all. On agar also, either there were no colonies or normal large colonies with occasional smaller ones were detected (data not shown). This indicates that the bacteria that survived ampicillin or nalidixic acid treatment cannot be distinguished from the normal cells which are consistent with the findings of Keren *et al.* (2004).

The difference between kanamycin and persisters of other antibiotics was also evident from the zone of inhibition by antibiotics (Fig. 2.A and B). The area immediately outside the zone of inhibition by kanamycin consisted of only small colonies and the size of colonies gradually increased as the antibiotic concentration decreased. With ampicillin, tetracycline and nalidixic acid, the zone of inhibition consisted of both large and small colonies (Fig. 2.B shows persisters of ampicillin; the same pattern was seen with tetracycline and nalidixic acid). This study indicates that among the four antibiotics, only kanamycin can produce a pure culture of persisters. Persisters isolated by kanamycin therefore exhibit a heritable phenotype which was not observed with persisters isolated with other antibiotics. Since other antibiotics cannot select a pure culture of persisters, the 'phenotypic shift' exhibited by these persisters was only due to the presence of some normal fast-dividing cells. The selection of persisters and their slow growth can partly explain the long post-antibiotic effect (the period of time after removal of antibiotics when bacterial

Table 5. Growth characteristics of persisters of *E.coli* DH-5α isolated by different antibiotics

Antibiotic*	Concn (ug/ml)	O.D. of bacterial culture						Colony growth (h)△
		In presence of antibiotics			On removal of antibiotics§			
		9h	12h	24h	9h	12h	24h	
TET	0	t	t	t	t	t	-	>15
	5	0.43±0.08	0.4±0.08	0.38±0.05	t	t	-	>15
	10	0.23±0.02	0.2±0.01	0.21±0.02	0.83±0.1	t	-	>15
	15	0.13±0.02	0.11±0.01	0.12±0.02	0.57±0.05	t	-	>15
	20	0.1±0.01	0.1±0.01	0.1±0.01	0.17±0.01	t	-	>15
AMP	0	t	t	t	t	t	t	>15
	2	0.33±0.03(L)	t(L)	t(L)	t	t	t	>15
	4	0.26±0.02(L)	0.74±0.12	t(L)	t	t	t	>15
	6	L	0.5±0.06	t(L)	t	t	t	>15
	8	L	L	t(L)	t	t	t	>15
	10	L	L	t(L)	t	t	t	>15
	12	L	L	L	1.08±0.15	t	t	>15
	16	L	L	L	0.34±0.04	t	t	>15
	20	L	L	L	>0.1	>0.1	>0.1	-
	25	L	L	L	>0.1	>0.1	>0.1	-
NAL	0	t	t	t	t	t	-	>15
	5	0.28±0.05(L)	t(L)	t(L)	t	t	-	>15
	10	0.16±0.03(L)	t(L)	t(L)	t	t	-	>15
	15	L	L	0.22±0.03(L)	t	t	-	>15
	20	L	L	L	0.37±0.05	t	-	>15
	25	L	L	L	0.44±0.08	t	-	>15

Results are shown as mean ± s.e.m.; $n=3$ for all antibiotics.

t – turbid ; L – lysis ; t(L) – turbid with lysis

* TET-tetracycline; AMP-ampicillin; NAL-nalidixic acid

§ persisters isolated were washed twice and incubated in fresh LB medium for 48 h.

△Time taken to form visible colonies after antibiotic-free persister cultures were plated on agar.

growth is not observed) exhibited by aminoglycosides (Gilbert 1991; Kapusnik *et al.* 1988). The *in vitro* post-antibiotic effect of kanamycin against *E. coli* DH5-α can be much longer, provided only a pure culture of extreme persisters remains.

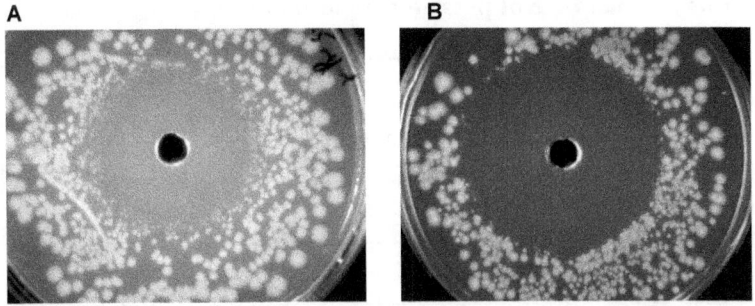

Fig.2. Aminoglycosides select persisters with different growth rates. A. Area at the zone of inhibition after kanamycin treatment consists of only small colonies. Larger colonies are seen radiating out as the antibiotic concentration decreases. B. Persisters remaining after ampicillin treatment. The zone of inhibition is abrupt and has both large and small colonies.

Tolerance of persisters to antibiotics

Persisters are reported to be responsible for recalcitrant infections and can tolerate high antibiotic concentrations (Lewis 2007).The ability of different persister groups to tolerate higher kanamycin concentrations was tested by treating each group of the persisters. Each group of the persisters was treated with the same concentration of kanamycin at which they were isolated and with higher concentrations. Persisters showed increased tolerance to kanamycin which is evident from the high MIC value (Table 6). All persisters were completely tolerant to the concentration at which they were isolated; that is, there was no significant difference between the growth rate of K-40 grown in antibiotic-free medium and those grown in a kanamycin concentration of 40 µg/ml (data not shown). However, with increasing concentrations, they became more and more sensitive. Depending on the type of persisters, an increased MIC up to 2-to-7 fold was noticed.

Antibiotic tolerance exhibited by persisters may be clinically significant because a sub lethal concentration of an aminoglycoside can result in the selection of persisters which may cause a latent infection later and will be difficult to eradicate since a high concentration of antibiotic needs to be used. In this aspect, persisters are similar to small colony variants (SCVs), variants of bacteria that grow slowly, form small colonies and are tolerant to antibiotics, especially aminoglycosides (Proctor *et al.* 1995; Proctor *et al.* 2006). Most of the

Table 6. Tolerance of persisters to antibiotics

Type of cell	Minimum inhibitory concentration[§]		
	Kanamycin (µg/ml)	Ampicillin (µg/ml)	Nalidixic acid(µg/ml)
con	53.3 ± 3.3	10.67± 0.7	14.3 ± 0.3
K-10	110 ± 5.8	10.4±0.5	14.7 ± 0.7
K-20	170 ± 14.6	11.3 ± 1.3	14.6 ± 0.3
K-30	283.3 ± 8.8	23.3 ± 1.7	24.7 ± 2
K-40	346.7 ± 26.1	28.7 ± 1.7	32.7 ± 5.5

Results are shown as mean ± s.e.m.; $n = 3$

[§] Minimum concentration of antibiotic that inhibited the visible growth of test organism after 48 h of incubation. MIC is usually determined after overnight incubation. Since persisters require more time for growth, MIC of all antibiotics was determined after 48 h of incubation.

SCVs reported are mutants of *Staphylococcus aureus* isolated from clinical infections and are auxotrophic to hemin, thiamine or thymidine (Proctor *et al.* 2006; von Eiff *et al.* 1997a, b). *E. coli* lack the ability to take up hemin (Roggenkamp *et al.* 1998; Sasarman *et al.* 1968) and the persisters of *E. coli* did not revert at any of the five different concentrations of thiamine or thymidine (data not shown). Moreover, persisters comprise a number of distinct subgroups with different growth characteristics, which contradicts the possibility of their being mutants, as the mutants reported earlier are single subpopulations. Aminoglycoside tolerance exhibited by a majority of SCVs is due to defects in electron transport resulting in reduced transmembrane potential that does not favor the uptake of aminoglycosides (Proctor *et al.* 2006). Aminoglycoside uptake and its killing rate are directly dependent on transmembrane potential (Eisenberg *et al.* 1984). Whereas SCVs may arise from genetic mutations (Proctor *et al.* 2006), persisters may have defects in electron transport other than mutations in the electron transport chain pathway. SCVs that are auxotrophic for a number of agents other than hemin, thiamine or thymidine as well as stable SCVs that are not auxotrophic to any of these agents, are also reported (Proctor *et al.* 2006). Since persisters form colonies of all sizes and were noticed with all bacterial cultures tested (Table 1., Fig. 1, Fig. 2.A.) and showed no reversion on addition of thiamine or thymidine, it can be assumed that they are naturally occurring forms. Moreover, after eliminating extreme persisters by repeatedly growing them in the early exponential phase, all of them reappeared after overnight incubation (Table 3). This raises the possibility that they arise from a natural process rather than from mutations and that these stable SCVs may be different from the mutants reported earlier. It is not documented whether electron transport defective SCVs also form colonies of various sizes. In such cases, the term 'small colony variants' itself becomes a misnomer as these variants can produce larger colonies also (the colony size of K-10 was indistinguishable from control after 36-40 h).

Aging can reduce the transmembrane potential since the expression of the electron transport chain pathway decreases with age and is common to humans, mice and flies (Zahn *et al.* 2006). Whether this is true for bacteria is not known. However, bacteria which were considered to be functionally immortal may also undergo aging and death (Ackermann *et al.* 2003; Stewart *et al.* 2005; Liu 1999). Stewart *et. al* (2005) studied the replicative senescence of *E. coli* and found that bacteria divide asymmetrically. They observed a rate of decline of growth by 1% in cells that inherited old poles. The cells growing slower were the ones that have more often inherited old poles. Persisters are not mutants; they comprise many subgroups that can be isolated from all cultures used and show a gradual reduction in growth rate and colony sizes which cannot be reverted back. It is therefore hypothesized that persisters are bacteria at different stages of aging which are tolerant to aminoglycosides due to their reduced uptake of the antibiotic resulting from a lower expression of the electron transport chain pathway (Fig. 4.B shows some differences in the protein expression profile of the control and persisters).

As per the model for bacterial aging proposed by Ackermann *et al.* (2003), Stewart *et al.* (2005) and Liu (1999), when a bacterium divides, the mother cell becomes older whereas the daughter cell is a rejuvenated young offspring that has a growth rate similar to normal cells (Fig. 3, model 2). However, the model proposed here is different (Fig. 3, model 3) because, with bacterial reproduction, the mother cell becomes older but the daughter cell is not a rejuvenated offspring and instead has the same age as the mother cell. The rejuvenated offspring hypothesis is not supported because persisters never attain the growth rate of normal cells even after they were passaged 3-4 times. Rather, they maintained their own growth rate which was slower than that of the normal cells.

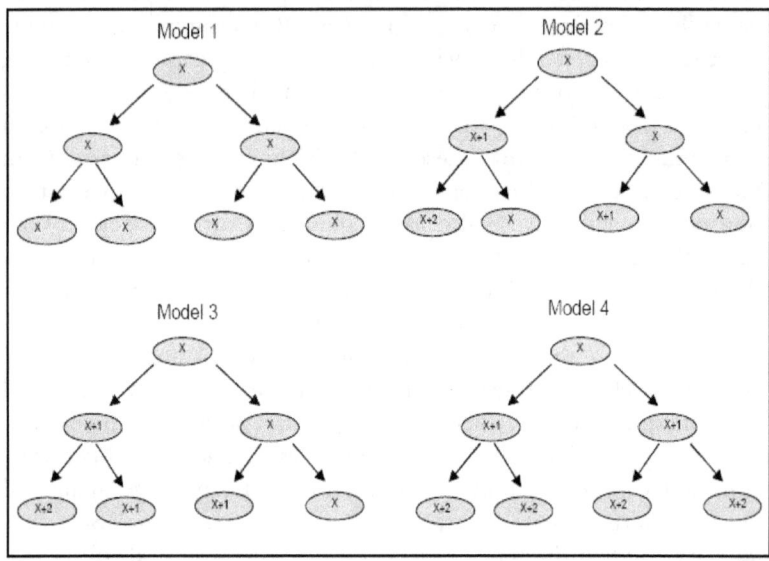

Fig.3. Models of bacterial aging. Model 1 assumes that bacteria are functionally immortal and there is no aging in bacteria. Here 'X' is considered the starting age of the bacterium. Model 2. As per this model, bacteria undergo aging. When a bacterium reproduces, the mother cell becomes older and shows a slower growth rate, but the daughter cell is a rejuvenated offspring which has the same growth rate as the normal cell. In this case, bacterial lineage is never lost. Due to the production of rejuvenated off-spring, a culture of aging bacteria will eventually have a growth rate as fast as normal cells. Model 3. proposes that as a bacterium reproduces, the mother cell becomes older whereas the daughter cell maintains the same age as the mother. Here also there is no loss of bacterial lineage. However, a culture of aging bacteria can never attain the same growth rate as normal since the daughter cell is not a rejuvenated, fast-growing offspring. This model is supported by the findings on persisters here. Model 4. As a bacterium divides, both mother and daughter cells become older. Here there will be a complete loss of bacterial lineage after a number of divisions. Even though this model is not supported here, it can not be rejected based on the results on persisters.

To test whether kanamycin tolerant persisters also exhibit tolerance to other antibiotics, persisters isolated using kanamycin were treated with different concentrations of ampicillin and nalidixic acid. While there was no significant difference among the control, K-10 and K-20 groups as far as MIC was considered, K-30 and K-40 did show increased MIC up to 2-to-3 fold (Table 6). The result demonstrates that persisters isolated by kanamycin may offer some tolerance to other antibiotics also, but not to the extent reported earlier (Li and Zhang 2007).

Persisters generate noise in protein expression
Individual cells of a genetically identical homogenous population of bacteria may show different protein expression levels referred to as noise in protein expression (Bar-Even et al. 2006; Elowitz et al. 2002; Ozbudak et al. 2002; Rao et al. 2002). Noise can be extrinsic when protein expression levels differ between individual cells of a genetically identical homogenous population or intrinsic when the differences arise due to inherent stochasticity of individual cells with proteins produced in random bursts (Cai et al. 2006, Swain et al. 2002). Sources of noise can be multiple such as variation in cell cycle stage, aging, epigenetic regulation, unequal segregation of mitochondria during cell division, fluctuations in upstream signaling, subtle differences in surrounding environments etc (Kaern et al. 2005; Swain et al. 2002).

To determine whether persisters are responsible for generation of noise, the expression levels of green fluorescent protein (GFP), an unnecessary protein for the bacteria, in normal cells and in persisters were measured. BL21 (DE-3) cells with plasmid containing the gene for GFP and a high basal level of GFP expression were used for this purpose. Since the presence of stressors like antibiotics, especially kanamycin, that inhibit protein synthesis may affect protein expression, GFP expression was measured in persisters grown in antibiotic-free medium. For this purpose, persisters of BL21 (DE-3) cells were isolated first, followed by washing to remove the antibiotics and further incubated in fresh medium without antibiotics. This was repeated once again to eliminate any effects of antibiotics in protein expression. GFP expression by persisters was then determined by Western blot which showed a decreasing gradient with persisters isolated using higher kanamycin concentrations (Fig. 4.A). It is predictable that there will be innumerable subgroups with subtle differences in protein expression levels. The difference in the protein

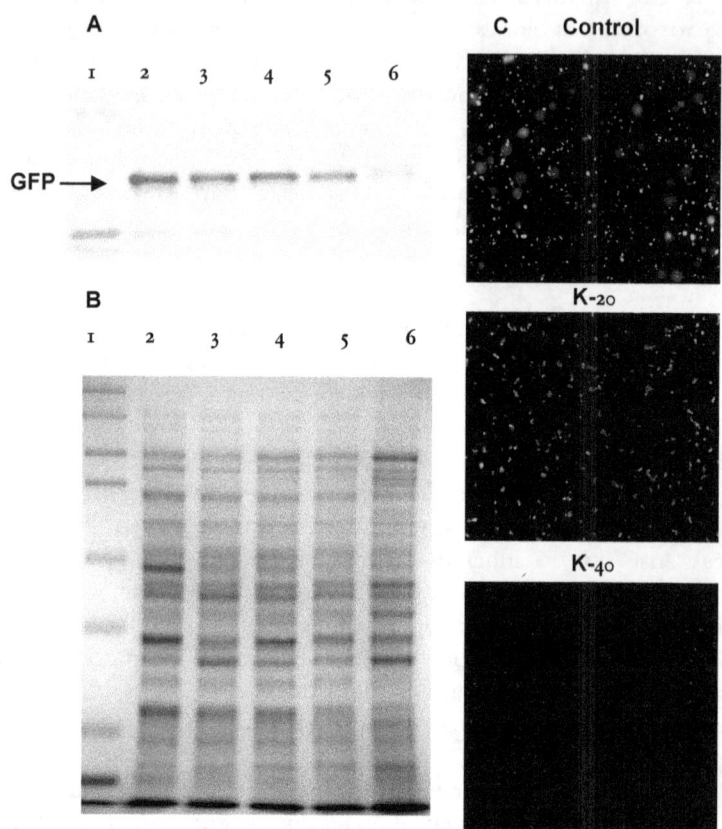

Fig.4. Persisters generate noise in protein expression. A. Western blot showing GFP expression by normal cells and persisters of BL21 (DE-3) cells. Lane 1: protein marker; lane 2:control; lane 3: K-10; lane 4: K-20; Lane 5: K-30; and lane 6: K-40. B. Protein fractions of BL21 (DE-3) normal and persisters resolved on 12% SDS-PAGE and visualized by Coomassie staining. Lane1: protein marker; lane 2: control; lane 3: K-10; lane 4: K-20; Lane 5: K-30; and lane 6: K-40. C. Intensity of GFP fluorescence in normal, K-20 and K-40 persisters of BL21 (DE-3) by fluorescence microscopy.

expression is not due to stochastic variation but instead results from the existence of distinct persister populations. The noise was also evident from immunofluorescence (Fig. 4.C). The intensity of fluorescence by GFP decreased as its expression was reduced, which was consistent with western blot results. However, an entire range of intensity was not obtained as in western blot. No difference in intensity was noticed between K-10 and K-20 nor between K-30 and K-40. The difference in protein expression was also visible from the intensity of fluorescence of colonies in solid medium (data not shown). While control gave the strongest intensity, K-40 showed only very low intensity. Again, a decreasing gradient of fluorescence was noticed as in western blot.

Significance of persisters

Persisters are slow-growing bacteria present in a normally dividing bacterial population; they are neither mutant nor induced by antibiotics. The 'phenotypic shift'of persisters reported earlier can be due to faulty experimental set-up resulting from suboptimal conditions of bacterial killing. Persisters can be aging bacteria and can be selected by aminoglycosides due to the peculiar property of aminoglycoside uptake by bacteria, which depends on the transmembrane potential. They comprise a number of subgroups and are one of the sources of bacterial heterogeneity and noise in protein expression. Even though persisters generate phenotypic heterogeneity, they may not offer an appreciable survival advantage to the population, as they do not revert to the normal parent population. Persisters may be aging bacteria with reduced levels of protein expression and thus may provide an excellent model for bacterial aging. The hypothesis that a 'true VBNC' state results from bacterial senescence is also supported here. The existence of many subgroups of persisters with different growth rates, colony sizes, antibiotic tolerance and protein expression levels warrants revision of some of the fundamental concepts in microbiology, including colony-forming units, stationary phase physiology, VBNC, post-antibiotic effects and aging.

Acknowledgements: I thank Dr. Ian. H. Mather, Department of Animal Sciences, University of Maryland, College Park, for providing lab facilities, chemicals and bacterial strains; Dr. Iqbal Hamza, University of Maryland, College Park, for providing DH-5alpha with GFP construct, nalidixic acid and fluorescent microscopy and Jae-Kwang Jeong, University of Maryland, for helping with fluorescent microscopy.

Section 2
A New Model of Bacterial Aging Incorporating SCVs, VBNC, and Persisters

What can we learn?

1. A new model of bacterial aging

I define persisters as small subpopulations of bacteria that can survive lethal concentrations of antibiotics. They are not mutants, nor do they carry antibiotic-resistant genes. They grow slowly and hence form small colonies on agar plates. Their growth rate is unaffected by the presence of antibiotics and they do not revert to normal growth on transfer to fresh medium without antibiotics. Even though they are tolerant to antibiotics, they may not have much clinical significance, as they may not cause infections due to their low virulence. I also hypothesize that the 'true persisters' are senescent bacteria.

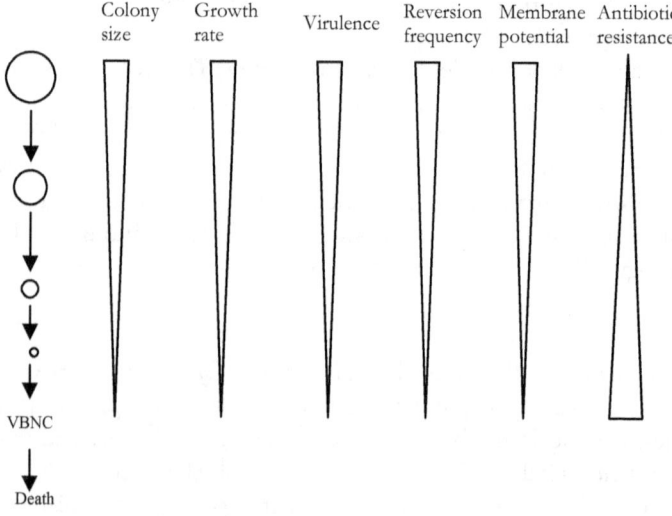

Fig. 1. Stages of replicative senescence and death in bacteria

Senescence and death are parts of the normal life cycle of bacteria (Fig. 1). Young bacteria divide faster and consequently form large colonies on agar. But as the bacteria undergo senescence, the rate of division gradually decreases, resulting in the formation of smaller colonies. Their membrane potential also reduces, which results in increased resistance to aminoglycoside antibiotics. Since they divide slowly, they become tolerant to other antibiotics also, as many antibiotics are less active against slow-dividing bacteria. Thus, senescent bacteria can persist in the presence of antibiotics.

If senescence is a part of the normal life cycle, it should be possible to isolate those bacteria at high frequencies. In fact, SCVs can be isolated from most bacterial species (Swingle 1935). I have isolated SCVs from *E.coli* DH5α, BL21 DE3, *Salmonella typhimurium* (Jacob 2007), *Pseudomonas putida* and *Burkholderia thailandensis* (unpublished; done at the University of Georgia). For bacteria exhibiting a high MIC against aminoglycosides (*E. coli* DH5α), a pure culture of SCVs can be isolated directly, by incubating bacterial cultures with high concentrations of the antibiotic (Jacob 2007). However, with those bacteria having a low MIC (e.g. *Pseudomonas putida*), it may be difficult to isolate pure cultures of SCVs from a parent population directly as the range of concentration needed to isolate them is very narrow. However, in such cases, SCVs can be isolated by incubating the bacterial culture with aminoglycosides at sub-MBC concentrations for 6-10 h, followed by a second round of treatment. This would help to eliminate most of the normal bacteria resulting in the selective growth of slow dividing bacteria. It may be difficult to obtain a pure culture of SCVs; however, colonies of many different sizes can be isolated from which stable SCVs can be selected.

Similarly, *E. coli* DH5α SCVs can be isolated not only from normal large-sized colonies, but also from smaller colonies as well. Small colonies of *E. coli* DH5α isolated using kanamycin at concentrations of 20 or 30 μg/ml (Jacob 2007) can be used to isolate still smaller colonies by incubating the bacteria at higher kanamycin concentrations. In a parent culture, the majority of the population in the stationary phase is the normal fast-dividing bacteria. The number of senescent bacteria is low as they are unable to divide further due to their growth disadvantage. However, after aminoglycoside treatment, fast-growing bacteria are selectively killed, leaving the senescent bacteria. These can further divide to give rise to more senescent ones that are more resistant to aminoglycosides due to the reduced antibiotic uptake resulting from their still lower membrane potential. Thus, by treating the small colonies with higher concentrations of aminoglycosides, much smaller colonies that cannot be isolated from a parent culture can be obtained (Fig. 2). For example, at 60 μg/ml of kanamycin, no growth of *E. coli* DH5α was noticed (Jacob 2007). However, at 50 μg/ml, SCVs could be isolated which, on further growth, were resistant to even 100 μg/ml of kanamycin. This means that the SCVs isolated at a kanamycin concentration of 50 μg/ml had given rise to new subpopulations that form still smaller colonies (lower growth rate) and exhibit more tolerance to kanamycin.

At stationary phase

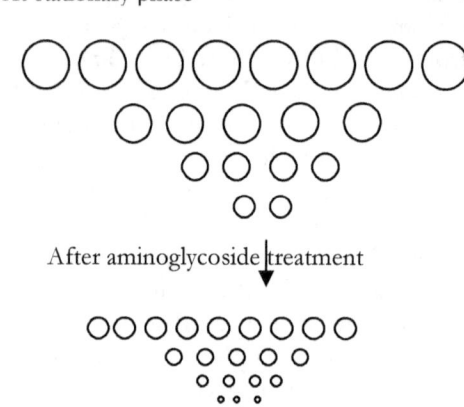

After aminoglycoside treatment

Fig.2. Different subpopulations of SCVs.

Moreover, even after removing the slow-dividing bacteria from a population by repeatedly growing the bacterial culture in the early exponential phase, SCVs reappear on reaching the stationary phase (Jacob 2007). Thus, SCVs can be isolated from any bacterial cultures that can form distinct colonies on agar, indicating that they are universal and are part of the normal life cycle of bacteria.

It is reported that SCVs are electron-deficient mutants which can revert to normal forms in the presence of auxotrophic agents (Proctor *et al.* 2006). However, isolation of many SCVs which are not auxotrophic has been reported (reviewed previously). Similarly, SCVs of *E. coli* DH5α, BL21 DE3 and *S. typhimurium* did not revert in the presence of these agents (Jacob 2007). It can be argued that these are hemin-deficient mutants that cannot revert in the presence of hemin due to their inability to uptake it. If they are indeed mutants, what is the advantage of having a high frequency of hemin-deficient mutants in a population when they do not have the ability to revert in the presence of hemin? To revert to normal forms, they require a second independent mutation that helps to uptake hemin, the frequency of which is usually very low (Roggenkamp *et al.* 2004).

Similarly, the mutant theory cannot explain why SCVs of different sizes can be isolated. By definition, SCVs are colonies that are smaller than one-tenth the size of normal colonies (Proctor *et al.* 2006). However, colonies much smaller than one-tenth size can also be isolated, and the molecular basis for such differences has not been explained. Also, as noted above, when small colonies are treated with higher concentrations of aminoglycoside, still smaller colonies can be isolated that are more tolerant to aminoglycosides. One can argue that these are also double mutants, but the frequency of these colonies is much too high to be double mutants. Thus, even though electron-deficient mutants can be frequently isolated both *in vitro* and *in vivo*, only bacterial senescence, and not the mutant theory, can explain the presence of SCVs of different sizes and their increased resistance to aminoglycosides. I propose that two groups of SCVs exist: electron-deficient mutants and non-mutants. The latter are normal senescent bacteria that are tolerant to antibiotics due to their reduced membrane potential and exhibit slow growth. Both of them may not have much clinical significance, even though the electron-deficient mutants can theoretically cause chronic infections as they have the ability to revert *in vitro* in the presence of auxotrophic agents. Their ability to revert *in vivo*, however, is questionable.

As per my proposal, the growth rate of bacteria (and hence the size of the colonies) reduces as bacteria undergo senescence and finally become non-culturable. At this stage, bacteria may remain viable and maintain membrane integrity, and they may be capable of a few more divisions (but not a sufficient number to form colonies). These non-culturable bacteria, however, will lose their viability and finally become morbid. The ability of senescent bacteria to revert depends on the stage in the life cycle. Towards the terminal stage of senescence, reversion to normal growth may not be possible. Thus, bacteria that form pinpoint colonies may not revert to normal growth. Beyond this stage, VBNCs also exist which cannot be resuscitated under any conditions. Thus, VBNCs are those bacteria nearing death that may remain viable and may have very limited capacity to divide but cannot be resuscitated to normal growth, thus representing another part of the bacterial life cycle (Fig. 1).

As per the current bacterial senescence model, mother cells on division undergo gradual aging, whereas the daughter cells formed from the mother cells are rejuvenated offspring (Stewart *et al.* 2005;

Ackermann *et al.* 2003). However, this may not be true towards the terminal stages of senescence (Jacob 2007) and hence the current model is incomplete. Older mother cells divide symmetrically and hence the daughter cells formed from such mother cells are not rejuvenated offspring (Jacob 2007).

The ability of 'early senescent' bacteria to revert to normal growth may depend on the ability of the mother cell to retain damaged proteins for itself. As the mother cell undergoes aging, it accumulates damaged macromolecules. Younger mother cells may segregate damaged macromolecules to themselves, thus giving rise to new rejuvenated daughter cells that can repopulate the culture. However, after many divisions, the mother cell loses its control over segregation and thereafter daughter cells may also inherit some damaged molecules. From this stage onwards, the ability of the colonies to revert gradually decreases and is finally lost completely. Thus, if we start with relatively young senescent bacteria (for example, the small colonies of *E. coli* DH5α cells isolated using a kanamycin concentration of 20-30 µg/ml (Jacob 2007)), the probability of reversion is higher due to the generation of rejuvenated daughter cells. On the other hand, the daughter cells derived from the terminal senescent bacteria or VBNCs may not revert at all since they also accumulate damaged proteins. Thus, my model of bacterial aging is similar to that of simple eukaryotic senescence (discussed earlier).

2. VBNC and SCVs share some common features.

If SCVs ultimately lead to the VBNC state, some common features between the two can be expected. VBNCs of many bacterial species share some common morphological features, including total size reduction (Chaiyanan *et al.* 2001; Zhong *et al.* 2009), loss of membrane integrity (Kusters *et al.* 1997; Chaiyanan *et al.* 2001; Alonso *et al.* 2002; Tangwatcharin *et al.* 2006; Zhong *et al.* 2009), formation of bleb-like structures (Kusters *et al.* 1997; Lazaro *et al.* 1999; Chen *et al.* 2009), increase in periplasmic space (Vattakaven *et al.* 2006; Zhong 2009), ghost cell formation and cell lysis (Phe *et al.* 2005; Vattakaven *et al.* 2006) and compressed nuclear regions (Chaiyanan 2001; Zhong *et al.* 2009). In the case of starvation, cells with reduced size and fewer division septa are noticed (Watson *et al.* 2006). Ghost cells, indicating cell lysis, can be seen as debris surrounding intact cells (Watson *et al.* 2006). The ultrastructure of SCVs also reveals ghost cells (Adler *et al.* 2005; Aoki *et al.* 1998; Wellinghausen *et al.* 2009). Similarly, SCVs with incomplete, branched and multiple cross walls without regular cell septation are also noticed (Proctor *et al.* 2006; Kahl *et al.* 2003; Wellinghausen *et al.* 2009). Thus, some common morphological features can be noticed between SCVs and VBNC, suggesting a relationship between the two. However, unlike VBNC, SCVs are reported to be larger than normal cells, probably due to impaired cell separation (Kahl *et al.* 2003; Seaman *et al.* 2007; Wellinghausen *et al.* 2009). These differences can be attributed to the fact that VBNCs are induced in all experiments through various stress conditions such as lower temperature, starvation etc., which ultimately lead to conditional senescence, whereas non-mutant SCVs have undergone replicative senescence.

3. Type-I and Type-II persisters do not represent true persisters.

Balaban *et al.* (2004) reported that persisters are of two types: Type-I and Type-II persisters. Type-I persisters are generated during the stationary phase and their switching rate during the exponential phase is negligible. Once transferred to fresh medium, they switch back to normal cells after remaining in a growth-arrested stage for many hours (Balaban *et al.* 2004; Gefen *et al.* 2008). Type-II persisters are generated continuously (Balaban *et al.* 2004; Gefen *et al.* 2008) and are formed through a phenotype-switching mechanism whereby a normal cell spontaneously becomes a Type-II persister and vice versa (Kussell *et al.* 2005). Thus both types of persisters have the ability to switch to normal cells even though the Type-I persisters take longer to exit the stationary phase. However, SCVs, which are naturally present in any bacterial culture, do not revert to normal growth in fresh medium. Even the electron deficient mutants require the auxotrophic agents to revert to normal growth. Thus, SCVs (electron deficient mutants or senescent bacteria) do not represent Type-I or Type-II persisters. In other words, Balaban *et al.* (2004) completely missed those slow-dividing bacteria. In fact, senescent bacteria should be considered to be the 'true persisters' as they are non-mutants and remain dormant in the presence of antibiotics.

4. The significance of SCVs: failure to achieve suboptimal PK/PD parameters.

Infections such as cystic fibrosis, osteomyelitis, tuberculosis etc., can persist in the body for a long time, even after prolonged antibiotic therapy. Both persisters and SCVs are implicated in these infections (Lewis 2007; Proctor *et al.* 2006). A common feature of such infections is the presence of large sequestered sites such as abscesses in bones or joints or the accumulation of pus in pleural cavity (Lew and Waldvogel 2004; Mogayzel and Flume 2010). For example, in CF patients, mutation in the CF transmembrane conductance regulator (CFTR) gene results in impaired chloride and water secretion, leading to viscous secretions in the airways (Mogayzel and Flume 2010). This affects mucociliary clearance, thus facilitating chronic bacterial infections. *P. aeruginosa* is one of the common bacteria responsible for chronic infections in CF patients (Mogayzel and Flume 2010; George *et al.* 2009). This is mainly due to the ability of the organism to adapt genetically to CF airways (Smith *et al.* 2006) and to survive in biofilms that develop in the viscous secretions in the airways (Singh *et al.* 2000). Similarly, chronic osteomyelitis is characterized by mild inflammation, presence of pockets of dead bone with abscess (sequestrum) and fistulous tracts (Lew and Waldvogel 2004). In the case of tuberculosis as well, bacteria exist in tuberculous granulomas, composed of epithelial macrophages, neutrophils and lymphocytes (Russell 2007). The centre of the granuloma can be necrotic and hypoxic and may contain dead macrophages and other cells (Russell 2007). Similarly, device-related infections are difficult to treat because they are caused by organisms that are protected by biofilms, resulting in increased antibiotic resistance (Stickler 2008). Thus, in all these cases, bacteria are exposed to different microenvironments and, depending on the location of the bacteria, the microenvironment induces differential responses to the bacteria, leading to changes in drug susceptibility.

Antibiotic treatment alone may not be sufficient for the treatment of chronic infections. This can be due to the restricted penetration of antibiotics at the site of infection due to the accumulation of pus or other materials and due to the formation of biofilms (Anderson and O'Toole 2008). For example, high doses of antibiotics may not be sufficient in the treatment of chronic osteomyelitis because of the limited circulation and poor penetration of antimicrobial agents into the site of infection (Parsons and Strauss 2004; Lew and Waldvogel 2004). It may not be possible to eradicate chronic osteomyelitis with antibiotic treatment alone. Debridement (surgical removal of dead bone tissues) and resection of infected bone and soft tissues along with lengthy antibiotic treatment are often required for successful treatment (Lew and Waldvogel 2004). Debridement may provide a viable vascularised environment and may help increase antibiotic penetration (Lew and Waldvogel 2004). Similarly, nutrient repletion and removal of airway obstruction along with antibiotic treatment are needed for the successful treatment of CF (Davis 2006). In device-related chronic infections, removal of the implants is recommended if they are the main source of infection (Seifert *et al.* 2003; von Eiff *et al.* 1999).

In chronic infections, the availability of total drugs at the site of infection can vary depending on the disease conditions. Due to the presence of large sequestered sites, the penetration of some antibiotics can be slow and hence the peak concentration at such sites is usually lower than in plasma, even though the total amount of drug penetrated may not be reduced, thus keeping the AUC unchanged (Donald 1997). Antibiotics such as fluoroquinolones, tetraclines or vancomycin can permeate rapidly into the biofilms, whereas aminoglycosides exhibit poor penetration (del Pozo and Patel 2007). For concentration-dependent antibiotics like aminoglycosides, where Cmax/MIC is the most important PK/PD parameter determining its efficacy, the slow penetration of antibiotics can result in reduced Cmax/MIC (Donald 1997), thus adversely affecting the efficacy of the drug. In large sequestered areas, this would result in the subinhibitory concentrations that would select the SCVs, whereas in other areas, peak concentrations may be too low to have an antibacterial effect. Thus, both normal and variant forms may survive depending on the location of bacteria. In fact, both forms are usually isolated from the clinical cases of chronic infections following treatment with aminoglycosides.

Even though SCVs may not cause infections, their isolation from the clinical samples may suggest that the antibiotic concentration may not be optimal at the site of the infection. Even though the level of antibiotic may be sufficient to achieve the optimal PK/PD parameters in plasma, it can be suboptimal at the site of infection. Isolation of SCVs from chronic infections had led the researchers to assume that SCVs were responsible for chronic infections (Proctor *et al.* 2006). However, SCVs cannot be considered the responsible agent of chronic infections if the normal forms co-cultured with SCVs are not the revertants of SCVs. Given the fact that normal forms are much more virulent and fast-growing than SCVs, it can be speculated that the normal forms are responsible for infections. Even though they are more susceptible to antibiotics than SCVs, the presence of sequestrated areas or biofilms may shield them from antibiotics and thus may help in their survival. Thus, researchers have given undue importance to SCVs while completely ignoring the role of normal virulent forms of bacteria.

It can be argued that normal bacteria that are not killed by antibiotics are the persisters (Lewis 2007). However, the persisters demonstrated *in vitro* and those surviving in biofilms are different, as previously discussed. The inability of antibiotics to kill 100% of bacteria *in vitro* is mostly an antibiotic property and depends on a number of technical factors that may not have much significance *in vivo*. Moreover, there are no indications that the persisters undergo a phenotypic shift and are responsible for either chronic infections or biofilm resistance.

5. Concentration-dependent killing of aminoglycosides can be visualized.

Aminoglycosides are concentration-dependent antibiotics whose efficacy is determined by Cmax/MIC value (Craig 1998). When kanamycin was allowed to diffuse through the agar plates, small colonies were seen at the zone of inhibition (Jacob 2007). As the amount of drug decreased due to less diffusion, the colony size also increased. This shows that the inhibition of bacterial growth is dependent on the concentration of aminoglycosides. Most of these colonies were not stable SCVs and hence are not senescent bacteria or electron-deficient mutants. Some of them, especially the pinpoint colonies, were dead (no growth if transferred to fresh medium), while others reverted to normal growth in fresh medium. Even though they were not stable SCVs, it shows that the bacterial killing by aminoglycosides is dependent on its concentration. With other antibiotics, both large and small colonies could be noticed. Thus, only aminoglycosides can be used to select a pure culture of slow-dividing bacteria.

6. Selection of slow-dividing bacteria is one of the reasons for the long PAE of aminoglycosides.

PAE is the period of time after the removal of an antibiotic when no bacteria growth is detected. It can be due to sublethal damage to the organism (den Hollander *et al.* 1998), the persistence of antibiotics retained in the cell (Stubbings *et al.* 2005; Stubbings *et al.* 2006) or the emergence of phenotypically resistant subpopulations (den Hollander *et al.* 1996). The duration of PAE is influenced by the type of antibiotics and organisms, concentration of antibiotics, total time of exposure, inoculum size and antibiotic combinations (Zhanel *et al.*

1991; Burgess 1999). Aminoglycosides exhibit a long PAE, and the selection of slow-dividing bacteria by aminoglycosides can be one of the reasons for its long PAE, apart from the above explanations. At concentrations below MBC, fast-growing bacteria are selectively killed, leaving mainly slow-growing bacteria, which are slower to re-grow. PAE of aminoglycosides can be increased by increasing the concentration of aminoglycosides (more slow growing bacteria are selected; hence growth takes longer time), increasing the incubation time (more 'perfect' killing of fast-dividing normal bacteria) and reducing the initial inoculum size (a high initial inoculum size may not kill all normal, fast-dividing bacteria, which will then repopulate the culture). On the other hand, antibiotics like penicillin exhibit low PAE. In this case, there is no selection of slow-growing bacteria; hence bacteria that survive are the normal population and show normal growth rates.

7. Bacterial senescence can generate noise in protein expression.

Noise in protein expression occurs due to differences among the individual cells of a genetically identical homogenous population (extrinsic noise) or due to inherent stochasticity of individual cells (intrinsic noise) resulting from the production of proteins in random bursts (Cai *et al.* 2006, Swain *et al.* 2002). For example, beta-galactosidase in individual cells is produced in a highly random and variable fashion, and the induction of the enzyme may not result in increased expression in all cells equally (Novick and Weiner 1957). Similarly, GFP expression also differs among cells in a population (Jacob 2007). One of the reasons for the different GFP expression among the cells is the presence of distinct subpopulations with subtle differences in protein expression levels (Jacob 2007). The existence of such subpopulations with different protein expression levels due to bacterial senescence has been proposed earlier (Stewart *et al.* 2005). It is hypothesized that these subpopulations are senescent bacteria of different age groups which exhibit different growth rates and protein expression levels (Jacob 2007).

Section 3
Senescent Bacteria as Potential Live Vaccines

This project was submitted to the Round 2 of Grand Challenges Explorations funded by Bill and Melinda Gates foundation. It was one of the projects selected by the reviewers amongst 3000 applications received by the Foundation, but failed to get the funding after the due diligent process.

OBJECTIVE: **To develop live bacterial vaccines using senescent bacteria, a naturally occurring avirulent subpopulation.**

Currently available bacterial vaccines that incorporate whole bacterium include live attenuated and killed/inactivated vaccines. Live attenuated vaccines can give a higher level of immune response than killed/inactivated vaccines as they can further multiply in the body increasing the antigen level. However, all bacteria cannot be attenuated using the traditional methods to be used as vaccines. Here a novel method for developing live bacterial vaccines is described which utilizes senescent bacteria, a naturally occurring avirulent subpopulation that does not require any artificial attenuation.

For nearly two hundred years, scientists considered bacteria as functionally immortal organisms. However, few reports published recently provide evidence that they also undergo aging and death (Stewart *et al.* 2005; Liu 1999). The models of bacterial aging are in incipient stage and at least two different models have evolved (Stewart *et al.* 2005; Liu 1999, Jacob 2007). A method to isolate senescent bacteria from a normally dividing bacterial population by exploiting their unique growth characteristics has been described (Jacob 2007). The isolation of senescent bacteria is based on the following principles:

1. As bacteria ages, they exhibit lower growth rate (Stewart *et al.* 2005; Liu 1999, Jacob 2007) and concurrently form smaller colonies (Jacob 2007)
2. Membrane potential (MP) of bacteria reduces as they undergo senescence due to the reduced expression of electron transport chain (ETC) pathway
3. Bacteria with low MP can be selectively isolated using aminoglycoside antibiotics

Aminoglycoside antibiotics are bactericidal drugs commonly used in the treatment of bacterial infections. The uptake of aminoglycosides into bacteria depends on their MP (Mates *et al.* 1982). In many species, MP reduces with age due to reduced expression of ETC pathway and seems to be a common signature of aging (Zahn *et al.* 2006). Even though lifespan varies greatly between species, the level of age regulation of ETC pathway is nearly the same (Zahn *et al.* 2006). It can be assumed that this is true in bacteria also. Young, active bacteria predictably have high MP resulting in a higher uptake of aminoglycoside leading to their rapid elimination whereas aging bacteria having low MP escape killing due to reduced uptake and are thus selectively grown. Thus, bacteria of different age groups can be isolated in a carefully controlled system of varying aminoglycoside concentrations which ensures the complete removal of normal fast growing bacteria leaving only the slow growing senescent bacteria (Jacob 2007). Mutants affecting the ETC may also appear along with aging bacteria. Most of these mutants are auxotrophic to some nutrients and can revert to the normal fast growing bacteria when grown in the presence of these auxotrophic agents (Proctor *et al.* 2006; von Eiff *et al.* 1997a, b). However, these mutants can be removed by repeatedly growing bacteria in early exponential phase of growth (Jacob 2007) followed by growing the bacterial culture in enriched media. Thus, even if mutants are present, they can be removed to ensure that a pure culture of senescent bacteria is obtained.

Senescent bacteria are less active and have slower growth rates (Stewart *et al.* 2005; Liu 1999, Jacob 2007). Their protein expression levels are also different and they may express only those proteins necessary for their survival. Expression of unnecessary proteins like green fluorescent protein (GFP) reduces as bacterium undergoes aging (Jacob 2007).Thus predominance shifts from virulence to survival. Moreover, they do not revert to virulent forms (Jacob 2007) as they accumulate oxidative damage in the natural process of aging. After a few more divisions, they gradually lose their viability and finally undergo death.

Currently available live attenuated vaccines use bacteria whose pathogenicity is weakened via attenuation by different methods. Senescent bacteria, on the other hand, are attenuated naturally during the process of aging. Vaccines incorporating senescent bacteria thus differ from the traditional live attenuated vaccines in using bacteria that do not require any artificial attenuation. 'Senescent bacterial vaccines' have the potential to become an effective and ideal group of vaccines because:

1. They may elicit good immune responses as they are live and are capable of multiplication.
2. Their virulence is, predictably, very low. They are slow dividing and exhibit reduced protein expression for virulence, electron transport chain pathway, GFP levels etc.
3. Reversion frequency may be low or absent as they are attenuated naturally by aging.
4. The use of naturally attenuated bacteria stimulates immune responses to antigen in their natural conformation itself and can be superior to the traditional methods of attenuation.

Other predictable advantages of 'senescent bacterial vaccines' include:

1. Easy to isolate senescent bacteria required for vaccine development
2. Vaccines can possibly be developed against most vaccine preventable diseases of bacterial origin as aging can be a universal phenomenon.
3. Prior knowledge on virulent proteins/genes not required.

Section II: Experimental plan

1. To test the 'universality' of bacterial senescence: Isolation of senescent bacteria from different strains of E.coli, Salmonella and Staphylococcus aureus of both ATCC cultures and clinical samples will be tested. These bacteria have high growth rates and are easy to cultivate under normal laboratory conditions. A common method for isolation of senescent bacteria from E.coli DH5α, BL21-DE3 and S. typhimurium LT2 has already been described (Jacob 2007). All normal colonies tested could give rise to many subpopulations of slow dividing senescent bacteria which immediately suggests the universal nature of bacterial senescence. However, Staphylococcus aureus, which is known to produce high frequency of electron transport deficient mutants and clinical samples of the above bacteria, need to be tested. If aging is a universal phenomenon, it should be possible to isolate senescent bacteria from any bacterial strains.

2. Pulse field gel electrophoresis will be done to confirm whether both the normal and senescent bacteria originated from the same clone of bacteria.

3. Ability of these subpopulations to revert in the presence of auxotrophic agents like hemin, menadione and thymidine will be tested. As per the hypothesis, senescent bacteria can not revert to original wild type population in the presence of any auxotrophic agents. Similarly, to rule out mutations as the cause for slow growth in senescent bacteria, selected genes of electron transport pathway will be amplified by polymerase chain reaction and sequenced.

4. To find a protein marker for aging bacteria: The protein profile of aging bacteria is different from normal bacteria (Jacob 2007). If a common protein marker could be identified for aging bacteria, it will be extremely useful in distinguishing senescent bacteria from the mutants.

5. Testing senescent bacteria *in vivo* in laboratory animals: As per the hypothesis, animals challenged with senescent bacteria will not get infection but will generate sufficient immune response so that a subsequent challenge with virulent bacteria will protect them from infection.

During the Phase I study, the emphasis will be given to prove the universal nature of bacterial aging. If aging is universal, it should be possible to isolate them from any strains used. Similarly, experiments should prove that senescent bacteria could be used as live vaccines that can provide sufficient protection against pathogenic bacteria without causing infection. If Phase I studies are successful, 'senescent bacterial vaccines' can be tested for further clinical trials.

References

Ackermann, M., Stearns, S. C., and Jenal, U. (2003). Senescence in a bacterium with asymmetric division. *Science* 300(5627), 1920.

Adler, H., Schraner, E. M., Frei, R. and Wild, P. (2005). Ultrastructureof a clinical isolate of *Staphylococcus aureus* small colony variant and its revertant. *Microsc Microanal* 11 (Suppl 2), 982-983.

Al-Fattani, M. A., and Douglas, L. J. (2004). Penetration of Candida biofilms by antifungal agents. *Antimicrob Agents Chemother* 48(9), 3291-7.

Alonso, J. L., Mascellaro, S., Moreno, Y., Ferrus, M. A., and Hernandez, J. (2002). Double-staining method for differentiation of morphological changes and membrane integrity of Campylobacter coli cells. *Applied and Environmental Microbiology* 68(10), 5151-5154.

Anderson, G. G., and O'Toole, G. A. (2008). Innate and induced resistance mechanisms of bacterial biofilms. *Bacterial Biofilms* 322, 85-105.

Aoki, Y., Yamauchi, Y., Hayashi, H., Takayama, Y. and Tsuji, A. (1998). Characterization of small colony variants of methicillin-resistant *Staphylococcus aureus* regrown in the presence of arbekacin *Journal of Infection and Chemotherapy* 4(3), 107-111.

Balaban, N. Q., Merrin, J., Chait, R., Kowalik, L., and Leibler, S. (2004). Bacterial persistence as a phenotypic switch. *Science* 305(5690), 1622-5.

Bar-Even, A., Paulsson, J., Maheshri, N., Carmi, M., O'Shea, E., Pilpel, Y., and Barkai, N. (2006). Noise in protein expression scales with natural protein abundance. *Nat Genet* 38(6), 636-43.

Bigger, J. W. (1944). Treatment of staphylococcal infections with penicillin. *Lancet ii*, 497-500.

Black, D. S., Kelly, A. J., Mardis, M. J., and Moyed, H. S. (1991). Structure and organization of hip, an operon that affects lethality due to inhibition of peptidoglycan or DNA synthesis. *J Bacteriol* 173(18), 5732-9.

Bogosian, G., and Bourneuf, E. V. (2001). A matter of bacterial life and death. *EMBO Rep* 2(9), 770-4.

Brooun, A., Liu, S., and Lewis, K. (2000). A dose-response study of antibiotic resistance in Pseudomonas aeruginosa biofilms. *Antimicrob Agents Chemother* 44(3), 640-6.

Burgess, D. S. (1999). Pharmacodynamic principles of antimicrobial therapy in the prevention of resistance. *Chest* 115(3), 19s-23s.

Cai, L., Friedman, N., and Xie, X. S. (2006). Stochastic protein expression in individual cells at the single molecule level. *Nature* 440(7082), 358-62.

Chaiyanan, S., Chaiyanan, S., Huq, A., Maugel, T., and Colwell, R. R. (2001). Viability of the Nonculturable Vibrio cholerae O1 and O139. *Systematic and Applied Microbiology* 24(3), 331-341.

Chen, S. Y., Jane, W. N., Chen, Y. S., and Wong, H. C. (2009). Morphological changes of Vibrio parahaemolyticus under cold and starvation stresses. *International Journal of Food Microbiology* 129(2), 157-165.

Costerton, J. W., Stewart, P. S., and Greenberg, E. P. (1999). Bacterial biofilms: a common cause of persistent infections. *Science* 284(5418), 1318-22.

Costerton, W., Veeh, R., Shirtliff, M., Pasmore, M., Post, C., and Ehrlich, G. (2003). The application of biofilm science to the study and control of chronic bacterial infections. *J Clin Invest* 112(10), 1466-77.

Craig, W. A. (1995). Interrelationship between Pharmacokinetics and Pharmacodynamics in Determining Dosage Regimens for Broad-Spectrum Cephalosporins. *Diagnostic Microbiology and Infectious Disease* 22(1-2), 89-96.

Craig, W. A. (1998). Pharmacokinetic/pharmacodynamic parameters: rationale for antibacterial dosing of mice and men. *Clin Infect Dis* 26(1), 1-10; quiz 11-2.

Craig, W. A., and Ebert, S. C. (1990). Killing and regrowth of bacteria *in vitro*: a review. *Scand J Infect Dis Suppl* 74, 63-70.

Davis, P. B. (2006). Cystic fibrosis since 1938. *American Journal of Respiratory and Critical Care Medicine* 173(5), 475-482.

del Pozo, J. L., and Patel, R. (2007). The challenge of treating biofilm-associated bacterial infection. *Clinical Pharmacology & Therapeutics* 82(2), 204-209.

den Hollander, J. G., Fuursted, K., Verbrugh, H. A., and Mouton, J. W. (1998). Duration and clinical relevance of postantibiotic effect in relation to the dosing interval. *Antimicrobial Agents and Chemotherapy* 42(4), 749-754.

Donald, P. R., Sirgel, F. A., Botha, F. J., Seifart, H. I., Parkin, D. P., Vandenplas, M. L., vandeWal, B. W., Maritz, J. S., and Mitchison, D. A. (1997). The early bactericidal activity of isoniazid related to its dose size in pulmonary tuberculosis. *American Journal of Respiratory and Critical Care Medicine* 156(3), 895-900.

Eisenberg, E. S., Mandel, L. J., Kaback, H. R., and Miller, M. H. (1984). Quantitative association between electrical potential across the cytoplasmic membrane and early gentamicin uptake and killing in Staphylococcus aureus. *J Bacteriol* 157(3), 863-7.

Elowitz, M. B., Levine, A. J., Siggia, E. D., and Swain, P. S. (2002). Stochastic gene expression in a single cell. *Science* 297(5584), 1183-6.

Freeman, C. D., Nicolau, D. P., Belliveau, P. P., and Nightingale, C. H. (1997). Once-daily dosing of aminoglycosides: review and recommendations for clinical practice. *J Antimicrob Chemother* 39(6), 677-86.

Gefen, O., Gabay, C., Mumcuoglu, M., Engel, G., and Balaban, N. Q. (2008). Single-cell protein induction dynamics reveals a period of vulnerability to antibiotics in persister bacteria. *Proceedings of the National Academy of Sciences of the United States of America* 105(16), 6145-6149.

George, A. M., Jones, P. M., and Middleton, P. G. (2009). Cystic fibrosis infections: treatment strategies and prospects. *FEMS Microbiol Lett* 300(2), 153-64.

Gilbert, D. N. (1991). Once-daily aminoglycoside therapy. *Antimicrob Agents Chemother* 35(3), 399-405.

Jacob, J. (2007). Persisters show heritable phenotype and generate bacterial heterogeneity and noise in protein expression. *Nature Precedings.* http://hdl.handle.net/10101/npre.2007.1411.2

Jacobs, M. R. (2001). Optimisation of antimicrobial therapy using pharmacokinetic and pharmacodynamic parameters. *Clin Microbiol Infect* 7(11), 589-96.

Kaern, M., Elston, T. C., Blake, W. J., and Collins, J. J. (2005). Stochasticity in gene expression: from theories to phenotypes. *Nat Rev Genet* 6(6), 451-64.

Kahl, B. C., Belling, G., Reichelt, R., Herrmann, M., Proctor, R. A., and Peters, G. (2003). Thymidine-dependent small-colony variants of Staphylococcus aureus exhibit gross morphological and ultrastructural changes consistent with impaired cell separation. *Journal of Clinical Microbiology* 41(1), 410-413.

Kapusnik, J. E., Hackbarth, C. J., Chambers, H. F., Carpenter, T., and Sande, M. A. (1988). Single, large, daily dosing versus intermittent dosing of tobramycin for treating experimental pseudomonas pneumonia. *J Infect Dis* 158(1), 7-12.

Keren, I., Kaldalu, N., Spoering, A., Wang, Y., and Lewis, K. (2004). Persister cells and tolerance to antimicrobials. *FEMS Microbiol Lett* 230(1), 13-8.

Korch, S. B., and Hill, T. M. (2006). Ectopic overexpression of wild-type and mutant hipA genes in Escherichia coli: effects on macromolecular synthesis and persister formation. *J Bacteriol* 188(11), 3826-36.

Kussell, E., Kishony, R., Balaban, N. Q., and Leibler, S. (2005). Bacterial persistence: A model of survival in changing environments. *Genetics* 169(4), 1807-1814.

Kusters, J. G., Gerrits, M. M., VanStrijp, J. A. G., and VandenbrouckeGrauls, C. M. J. E. (1997). Coccoid forms of Helicobacter pylori are the morphologic manifestation of cell death. *Infection and Immunity* 65(9), 3672-3679.

Lazaro, B., Carcamo, J., Audicana, A., Perales, I., and Fernandez-Astorga, A. (1999). Viability and DNA maintenance in nonculturable spiral Campylobacter jejuni cells after long-term exposure to low temperatures. *Applied and Environmental Microbiology* 65(10), 4677-4681.

Lew, D. P., and Waldvogel, F. A. (2004). Osteomyelitis. *Lancet* 364(9431), 369-379.

Lewis, K. (2007). Persister cells, dormancy and infectious disease. *Nat Rev Microbiol* 5(1), 48-56.

Li, Y., and Zhang, Y. (2007). PhoU is a persistence switch involved in persister formation and tolerance to multiple antibiotics and stresses in Escherichia coli. *Antimicrob Agents Chemother* 51(6), 2092-9.

Liu, S. V. (1999). Tracking bacterial growth in liquid medium and a new bacterial life model. *Science in China (Series C)* 42, 644-654.

Mates, S. M., Eisenberg, E. S., Mandel, L. J., Patel, L., Kaback, H. R., and Miller, M. H. (1982). Membrane-Potential and Gentamicin Uptake in Staphylococcus-Aureus. *Proceedings of the National Academy of Sciences of the United States of America-Biological Sciences* 79(21), 6693-6697.

Mogayzel, P. J., Jr., and Flume, P. A. (2010). Update in cystic fibrosis 2009. *Am J Respir Crit Care Med* 181(6), 539-44.

Moyed, H. S., and Bertrand, K. P. (1983). hipA, a newly recognized gene of Escherichia coli K-12 that affects frequency of persistence after inhibition of murein synthesis. *J Bacteriol* 155(2), 768-75.

Moyed, H. S., and Broderick, S. H. (1986). Molecular cloning and expression of hipA, a gene of Escherichia coli K-12 that affects frequency of persistence after inhibition of murein synthesis. *J Bacteriol* 166(2), 399-403.

Novick, A., and Weiner, M. (1957). Enzyme Induction as an All-or-None Phenomenon. *Proceedings of the National Academy of Sciences of the United States of America* 43(7), 553-566.

Nystrom, T. (2001). Not quite dead enough: on bacterial life, culturability, senescence, and death. *Arch Microbiol* 176(3), 159-64.

Oliver, J. D. (2005). The viable but nonculturable state in bacteria. *J Microbiol* 43 Spec No, 93-100.

Ozbudak, E. M., Thattai, M., Kurtser, I., Grossman, A. D., and van Oudenaarden, A. (2002). Regulation of noise in the expression of a single gene. *Nat Genet* 31(1), 69-73.

Parsons, B., and Strauss, E. (2004). Surgical management of chronic osteomyelitis. *American Journal of Surgery* 188(1A), 57s-66s.

Phe, M. H., Dossot, M., Guilloteau, H., and Block, J. C. (2005). Nucleic acid fluorochromes and flow cytometry prove useful in assessing the effect of chlorination on drinking water bacteria. *Water Research* 39(15), 3618-3628.

Proctor, R. A., van Langevelde, P., Kristjansson, M., Maslow, J. N., and Arbeit, R. D. (1995). Persistent and relapsing infections associated with small-colony variants of Staphylococcus aureus. *Clin Infect Dis* 20(1), 95-102.

Proctor, R. A., von Eiff, C., Kahl, B. C., Becker, K., McNamara, P., Herrmann, M., and Peters, G. (2006). Small colony variants: a pathogenic form of bacteria that facilitates persistent and recurrent infections. *Nat Rev Microbiol* 4(4), 295-305.

Rao, C. V., Wolf, D. M., and Arkin, A. P. (2002). Control, exploitation and tolerance of intracellular noise. *Nature* 420(6912), 231-7.

Roggenkamp, A., Sing, A., Hornef, M., Brunner, U., Autenrieth, I. B., and Heesemann, J. (1998). Chronic prosthetic hip infection caused by a small-colony variant of Escherichia coli. *J Clin Microbiol* 36(9), 2530-4.

Russell, D. G. (2007). Who puts the tubercle in tuberculosis? *Nature Reviews Microbiology* 5(1), 39-47.

Sasarman, A., Surdeanu, M., Szegli, G., Horodniceanu, T., Greceanu, V., and Dumitrescu, A. (1968). Hemin-deficient mutants of Escherichia coli K-12. *J Bacteriol* 96(2), 570-2.

Seaman, P. F., Ochs, D., and Day, M. J. (2007). Small-colony variants: a novel mechanism for triclosan resistance in methicillin-resistant Staphylococcus aureus. *Journal of Antimicrobial Chemotherapy* 59(1), 43-50.

Seifert, H., Wisplinghoff, H., Schnabel, P., and von Eiff, C. (2003). Small colony variants of Staphylococcus aureus and pacemaker-related infection. *Emerging Infectious Diseases* 9(10), 1316-1318.

Singh, P. K., Schaefer, A. L., Parsek, M. R., Moninger, T. O., Welsh, M. J., and Greenberg, E. P. (2000). Quorum-sensing signals indicate that cystic fibrosis lungs are infected with bacterial biofilms. *Nature* 407(6805), 762-764.

Smith, E. E., Buckley, D. G., Wu, Z., Saenphimmachak, C., Hoffman, L. R., D'Argenio, D. A., Miller, S. I., Ramsey, B. W., Speert, D. P., Moskowitz, S. M., Burns, J. L., Kaul, R., and Olson, M. V. (2006). Genetic adaptation by Pseudomonas aeruginosa to the airways of cystic fibrosis patients. *Proc Natl Acad Sci U S A* 103(22), 8487-92.

Spoering, A. L., and Lewis, K. (2001). Biofilms and planktonic cells of Pseudomonas aeruginosa have similar resistance to killing by antimicrobials. *J Bacteriol* 183(23), 6746-51.

Stewart, E. J., Madden, R., Paul, G., and Taddei, F. (2005). Aging and death in an organism that reproduces by morphologically symmetric division. *PLoS Biol* 3(2), e45.

Stickler, D. J. (2008). Bacterial biofilms in patients with indwelling urinary catheters. *Nature Clinical Practice Urology* 5(11), 598-608.

Stubbings, W., Bostock, J., Ingham, E., and Chopra, I. (2005). Deletion of the multiple-drug efflux pump AcrAB in Escherichia coli prolongs the postantibiotic effect. *Antimicrobial Agents and Chemotherapy* 49(3), 1206-1208.

Stubbings, W., Bostock, J., Ingham, E., and Chopra, I. (2006). Mechanisms of the post-antibiotic effects induced by rifampicin and gentamicin in Escherichia coli. *Journal of Antimicrobial Chemotherapy* 58(2), 444-448.

Swain, P. S., Elowitz, M. B., and Siggia, E. D. (2002). Intrinsic and extrinsic contributions to stochasticity in gene expression. *Proc Natl Acad Sci U S A* 99(20), 12795-800.

Swingle, E. L. (1935). Studies on Small Colony Variants of Staphylococcus aureus. *J Bacteriol* 29(5), 467-89.

Tam, V. H., Kabbara, S., Vo, G., Schilling, A. N., and Coyle, E. A. (2006). Comparative pharmacodynamics of gentamicin against Staphylococcus aureus and Pseudomonas aeruginosa. *Antimicrob Agents Chemother* 50(8), 2626-31.

Tangwatcharin, P., Chanthachum, S., Khopaibool, P., and Griffiths, M. W. (2006). Morphological and physiological responses of Campylobacter jejuni to stress. *Journal of Food Protection* 69(11), 2747-2753.

Vattakaven, T., Bond, P., Bradley, G., and Munn, C. B. (2006). Differential effects of temperature and starvation on induction of the viable-but-nonculturable state in the coral pathogens Vibrio shiloi and Vibrio tasmaniensis. *Applied and Environmental Microbiology* 72(10), 6508-6513.

von Eiff, C., Bettin, D., Proctor, R. A., Rolauffs, B., Lindner, N., Winkelmann, W., and Peters, G. (1997a). Recovery of small colony variants of Staphylococcus aureus following gentamicin bead placement for osteomyelitis. *Clin Infect Dis* 25(5), 1250-1.

von Eiff, C., Heilmann, C., Proctor, R. A., Woltz, C., Peters, G., and Gotz, F. (1997b). A site-directed Staphylococcus aureus hemB mutant is a small-colony variant which persists intracellularly. *J Bacteriol* 179(15), 4706-12.

von Eiff, C., Vaudaux, P., Kahl, B. C., Lew, D., Emler, S., Schmidt, A., Peters, G., and Proctor, R. A. (1999). Bloodstream infections caused by small-colony variants of coagulase-negative staphylococci following pacemaker implantation. *Clinical Infectious Diseases* 29(4), 932-934.

Watson, S. P., Clements, M. O., and Foster, S. J. (1998). Characterization of the starvation-survival response of Staphylococcus aureus. *Journal of Bacteriology* 180(7), 1750-1758.

Wellinghausen, N., Chatterjee, I., Berger, A., Niederfuehr, A., Proctor, R. A., and Kahl, B. C. (2009). Characterization of Clinical Enterococcus faecalis Small-Colony Variants. *Journal of Clinical Microbiology* 47(9), 2802-2811.

Woolfrey, B. F., and Enright, M. A. (1990). Ampicillin killing curve patterns for ampicillin-susceptible nontypeable Haemophilus influenzae strains by the agar dilution plate count method. *Antimicrob Agents Chemother* 34(6), 1079-87.

Woolfrey, B. F., Gresser-Burns, M. E., and Lally, R. T. (1987). Ampicillin killing curve patterns of Haemophilus influenzae type b isolates by agar dilution plate count method. *Antimicrob Agents Chemother* 31(11), 1711-7.

Zahn, J. M., Sonu, R., Vogel, H., Crane, E., Mazan-Mamczarz, K., Rabkin, R., Davis, R. W., Becker, K. G., Owen, A. B., and Kim, S. K. (2006). Transcriptional profiling of aging in human muscle reveals a common aging signature. *PLoS Genet* 2(7), e115.

Zhanel, G. G., Hoban, D. J., and Harding, G. K. M. (1991). The Postantibiotic Effect - a Review of Invitro and Invivo Data. *Dicp-the Annals of Pharmacotherapy* 25(2), 153-163.

Zhong, L., Chen, J., Zhang, X. H., and Jiang, Y. A. (2009). Entry of Vibrio cincinnatiensis into viable but nonculturable state and its resuscitation. *Letters in Applied Microbiology* 48(2), 247-252.

CHAPTER VII

CONCLUSIONS AND PREDICTIONS

Conclusions

1. The *in vitro* inoculum effect results from the destruction of the antibiotic by enzymes, the decrease in antibiotic concentration per cell or the loss of biological activity of the drug by antibiotic trapping or binding of antibiotics to cell wall components. The inoculum effect may not have much significance in clinical settings where the dosing regimens are repeated for days. Hence conventional MIC values, based on a standard inoculum size, may be sufficient for the PK/PD assessment of antibiotics. However, at high inoculum, the chances of selection of mutants are also high, if MIC values obtained from the standard inoculum are used. In order to prevent the selective growth of mutants, it may be preferable to maintain the concentration of antibiotics above the MPC (mutant prevention concentration). Thus, inoculum size can become an important factor if the aim is to prevent the growth of mutants.

2. Phenotypic shift of persisters is only an *in vitro* illusion. The phenotypic shift demonstrated in many experiments is the result of suboptimal conditions of bacterial killing provided in the experiments and is related to the inoculum effect. Due to the inoculum effect, the survivors may remain dormant in the presence of sub-inhibitory antibiotic concentrations. However, this survival may not have significance in clinical settings where the dosing regimens are repeated. Hence persisters isolated *in vitro* after a single dose of antibiotic cannot be considered responsible for chronic or recurrent infections.

3. 100% *in vitro* bacterial killing is not an absolute necessity for successful treatment of bacterial infections *in vivo*.

4. hipA may not have a specific role in persister formation. Rather it can be one of the several proteins involved in general bacterial stress responses.

5. 'True persisters' are senescent bacteria. Due to their slow growth and inability to give rise to rejuvenated offspring, they form small colonies on agar. However, they may not have clinical significance even though they are tolerant to antibiotics. Type-I or Type-II persisters do not represent these true persisters.

6. SCVs are not a single subpopulation but rather comprise colonies of different sizes. Hence the current definition for SCVs is not appropriate. SCV phenotypes can differ depending on the colony size, reversion frequency and the method of selection. Similarly, all SCVs are not hemin, thiamine, menadione or thymidine auxotrophs.

7. Even though hemin or menadione mutants revert *in vitro*, their ability to revert *in vivo* is not known. Similarly, the condition under which the reversion occurs in vivo is also not known. Unless they revert to normal forms *in vivo*, they can not be considered responsible for chronic infections.

8. SCVs may not be responsible for chronic or recurrent infections, nor can they cause fatal infections. Rather, they may be selected by antibiotic pressure and remain in the tissues without causing infections until they finally die. Normal bacteria, co-cultured with SCVs, may be responsible for chronic infections.

9. Isolation of SCVs from clinical samples of chronic infections indicates that the antibiotic treatment may not be optimal at the site of infection, even though it may achieve the optimal PK/PD parameters in plasma.

10. A switching mechanism between the normal bacterial population and SCVs may not exist. Under antibiotic pressure, some bacteria may reduce their growth rate or undergo repair. These bacteria may form small colonies on agar and revert to normal growth once transferred to antibiotic-free medium. This reversion can not be considered a switching mechanism.

11. Selection of slow-dividing bacteria is one of the reasons for the long PAE of aminoglycosides.

12. VBNC may not be a successful phenotype and hence resuscitation of VBNC may not be responsible for the seasonal outbreak of cholera in endemic areas.

13. Since *Vibrio* species are regularly introduced into the aquatic environment, autochthonous *V. cholerae* may not be responsible for the presence of pandemic strains of cholera across the world. Rather, the pandemic strains might have originated from a sin-

gle/few area(s) and then spread to different parts of the world. Human-associated sources of infection can be very important in the spread of disease.

14. The current model of bacterial aging is not complete. Only young mother cells may have the ability to give rise to rejuvenated offspring. Old mother cells may not give rise to rejuvenated offspring and hence the whole lineage arising from an old mother cell may undergo senescence. Thus, young mother cells undergo asymmetrical division, whereas older mother cells undergo more symmetrical division. In this aspect, the senescence model of *E. coli* closely resembles that of the simple eukaryotes like yeast.

15. Prolonged stress conditions may result in conditional senescence in bacteria. The ability of bacteria to revert to normal growth depends on the stage of conditional senescence. In this case, bacteria may grow normally or not at all. On the other hand, aminoglycoside pressure selects bacteria that had undergone replicative senescence. Here too, the ability of bacteria to revert depends on the stage of senescence. Very old mother cells may not revert to normal growth, nor can they give rise to rejuvenated offspring.

16. One of the sources of noise in bacterial protein expression is the presence of subpopulations of slow-dividing senescent bacteria in the culture.

Predictions

Since it may not be possible for me to work on specific bacterial mutants, some predictions are made here based on my findings and hypothesis.

1. Protein expression patterns of persisters (Lewis 2007) and those bacteria exposed to sub-inhibitory concentrations of antibiotics will be almost the same.

   I had suggested that persisters (as demonstrated in many experiments) are those dormant bacteria exposed to sub-inhibitory concentrations of antibiotics. Hence the protein expression pattern of a small inoculum of bacterial culture exposed to sub-inhibitory antibiotic concentration will be almost similar to that of the persisters.

2. Infections caused by experimental inoculation of bacterial *hipA* mutants into animals can be treated successfully with the same dosage regimen used for the wild type, provided *hipA* mutants exhibit the same MIC as the wild type (MIC should be calculated as per NCCLS protocols).

It is assumed that *hipA* mutants are capable of causing chronic infections since they can not be killed at the MBC concentration that can kill the wild type bacteria (however, both exhibit the same MIC). I suggest that the antibiotic dosage regimen used for the successful treatment of the wild type bacteria will be sufficient for the mutants also, provided the *hipA* mutants exhibit the same MIC (calculated as per NCCLS protocols) as the wild type. This is because the PK/PD parameter that determines the efficacy of antibiotics is based on the MIC value. However, if the site of infection is inaccessible to the immune system, a different dosage regimen may be required.

3. *hemB* or *menD* mutants grown in the presence of hemin or mena-dione respectively will still produce SCVs.

When the mutants are grown in the presence of auxotrophic agents, they will revert to normal growth, which will ultimately generate senescent bacteria. These senescent bacteria may form small colonies but may not be auxotrophic to hemin or menadione.

4. It may be possible to use SCVs as vaccines since they grow slowly and are less virulent.

5. Senescent bacteria will be good models for epigenetic research.

6. SCVs of *E. coli* isolated using kanamycin will have increased amounts of altered proteins.

Since senescence is associated with the accumulation of altered proteins, especially carbonylated proteins, those SCVs will have high amounts of altered proteins.

7. Since senescence is associated with accumulation of deletions or mutations, SCVs of *E. coli* isolated using kanamycin may show multiple mutations. However, those mutations may not be specific like *hemB* or *menD* mutations.

CHAPTER VIII

CREATING ILLUSIONS...

On September 15, 2008, Lehman Brothers, a global financial services firm, filed for Chapter 11 bankruptcy protection and made history as the largest bankruptcy filing in the U.S. Many financial institutions collapsed like houses of cards. The world entered a recession, millions of jobs were lost and trillions of dollars were pumped into the economy to prevent another great depression. Even though the world averted another depression, it may take a few more years for the financial world to return to normal growth. As a result of the recession, many countries face a debt crisis. The U.S. federal deficit, as percentage of GDP, is estimated to be around 10% for fiscal year 2010, and the national debt of the U.S. stands at more than 12.5 trillion dollars today, with a projection of about 14 trillions (close to 100% of its GDP) by the end of 2010. In Iceland, the financial crisis led to the nationalization of all three of the country's major banks following the collapse of its currency. Greece is facing one of the most severe economic crises with a huge budget deficit and a high debt-to-GDP ratio. The European countries collectively known as PIIGS (Portugal, Italy, Ireland, Greece and Spain) are also facing severe unemployment and financial crises.

The fall of Lehman Brothers and the economic recession did not happen overnight, nor were they accidents. One of the immediate triggers of the crisis was the collapse of the global housing bubble. The housing bubble peaked in the U.S. in 2006, artificially inflating the prices of houses and real estate markets and reaching unsustainable levels relative to incomes. People who considered themselves rich based on the market value of houses and real estate soon faced the reality of illusion. Though the fall of Lehman Brothers came as a surprise to many, there were at least a couple of people who had correctly predicted the economic recession. One of them was Dr. Nouriel Roubini, Professor of Economics at the Stern School of Business in New York. As early as 2005, when the housing bubble was peaking, he predicted that the United States could face a housing bust and an economic recession. Although he was labeled as 'Dr. Doom' for his negative predictions, he was later proven to be accurate. Thus, the reason for the economic recession was not due to the lack of adequate information or warnings, but rather due to our failure to integrate different views and restrict the potential for some people to manipulate situations to their own advantage.

Bubbles exist not only in the financial world, but also in science. One of the main reasons for bubbles in scientific research is peer pressure. The practice of evaluating scientists based on performance indicators rather than assessing the research itself (Lawrence 2003) can lead to unhealthy competition which, when coupled with the pressure to publish in leading journals and to attract grants, can lead to research misconduct. In one survey, more than a third of U.S. scientists have admitted to misbehaving (Wadman 2005). Even though research misconduct usually includes falsification, fabrication, and plagiarism (FFP), it is also argued that the scientific community should look beyond FFP to a wider range of research misconduct (Martinson *et al.* 2005). Pressure to publish may also lead to worthless publications (de Carvalho 2006). One might argue that the rigorous process of peer-review may be sufficient to prevent flawed research publications. Unfortunately, this may not be true.

Why does peer-review miss those bubbles?

The peer review process is not without flaws. As noted by Gannon (2001), reviewers may not be competent to evaluate the topic under review and may have their own time constraints. It is possible that the reviewers are also misled by illusions created by the authors. A study by Richards (2007) indicated a lack of evidence in terms of the effect of referees' training and author concealment in regard to the quality of publications. Thus the chances of missing the bubbles or illusions by the reviewers are high. Once an article gets published in a peer-reviewed journal, the research findings proposed in the article are taken for granted as 'true findings'. Thus the initial finding becomes a scientific truth and eventually it becomes difficult to publish contradictory findings (DeCoursey 2006).

In this book I focused on how researchers have misinterpreted their data to convert relatively insignificant *in vitro* findings to topics of utmost public health importance. A common feature of all topics is the ability of bacteria to remain dormant under unfavorable conditions and subsequently 'shift', 'resuscitate' or 'revert' to normal forms during favorable conditions, resulting in chronic or recurrent infections. Since chronic bacterial infections in humans are always problematic, it is natural that these studies attract attention from the scientific community and from different funding sources. I have argued that, by closely scrutinizing a subject, it is possible to identify the flaws and thus save money and resources.

In the case of persisters, researchers have relied on *in vitro* tests to demonstrate the phenotypic shift of persisters. Using the knowledge that not all bacteria may be killed by antibiotics in biofilms and during chronic infections, researchers have proved the existence of persisters by simply demonstrating the presence of normal bacteria in a culture medium containing antibiotics. I assume that their aim was to prove the existence of persisters first and then propose that persisters are responsible for the antibiotic resistance of biofilms and the recalcitrance of chronic infections. I see the phenotypic shift of persisters as a 'plot' rather than as a scientific finding. These researchers have intentionally ignored the fundamentals of bacterial killing kinetics to avoid many questions. Without doing the necessary background work, data were published and hypotheses were made with the available data. There are many flaws in those experiments which can be easily identified by scientists working in the field of bacterial killing kinetics. However, due to the negligence of some journal reviewers, those flaws were not identified. Therefore, it has now become an "accepted scientific truth" that persisters are responsible for chronic infections. Millions of dollars of government funds have already been diverted to this field and millions more will be spent on this 'hot topic' in the future.

Regarding SCVs, researchers have isolated those variants from a variety of chronic infections and have proposed that the reversion of SCVs is responsible for many chronic infections. However, such reversion *in vivo* has not been proved. More importantly, in a majority of the cases, SCVs were co-cultured with normal forms. The researchers completely ignored the role of normal forms and gave importance only to SCVs. By demonstrating that both the forms are clonal (this point was stressed in most of the articles), they have artificially created an illusion that the normal forms are the revertants of SCVs. To give more importance and authenticity to SCVs, it was even proposed that SCVs were responsible for some fatal infections, the findings of which are highly questionable.

Twenty-eight years have passed since VBNCs were first proposed by Colwell's lab (Xu *et al.* 1982). But until now, it has not been conclusively proved that VBNCs are responsible for seasonal cholera outbreaks. Opponents argue that the regrowth of bacteria is only due to the presence of a few culturable bacteria and not due to the resuscitation of VBNCs. Interestingly, in 1994, Colwell's group admitted that the recovery of bacterial growth could be indeed due to the re-growth of a few culturable bacteria and not due to the resuscitation of VBNCs (Chowdhury *et al.* 1994; Ravel *et al.* 1995). However, even after publishing those articles, they maintained that VBNCs are important in outbreaks of diseases like cholera. If the VBNC concept were correct, its potential application would have been enormous. However, our knowledge about VBNCs has not grown except for the reports that more organisms are capable of entry to the VBNC state. Those who support the VBNC hypothesis had tried to downplay criticism by suggesting that the opposition was mainly against the existence of VBNCs or its nomenclature (Oliver 2005). However, the main question regarding VBNCs is whether they have the ability to resuscitate and cause disease outbreak. Moreover, in order to support the VBNC hypothesis, Colwell's group proposes that the autochthonous *V. cholerae* is responsible for the pandemic spread of cholera. Even though *V. cholerae* is autochthonous to marine and estuarine environments, the presence of carriers and human-associated activities may have significant roles in the spread of the disease. There is no convincing evidence to prove the role of autochthonous *V. cholerae* in the global spread of cholera. Thus, it is possible that Colwell's group might have created two illusions: one regarding the role of VBNCs in the seasonal outbreak of cholera, and the second regarding the role of autochthonous *V. cholerae* in spreading the disease across different continents.

Unlike the research findings on persisters, SCVs and VBNCs, findings on *E. coli* senescence do not involve a systemic misinterpretation of scientific data. However, the current knowledge on bacterial senescence may be incomplete, as discussed previously.

Impact of illusions

Even though the direct impact of research illusions on the economy may not be very large, bubbles can cause manifold damages. Illusions cause the wasting of public money along with time, energy and resources. And more importantly, the future scientists who get attracted to these fields will also be misled, creating a vicious cycle that will result in more wasting of resources. In this period of economic hardship, researchers around the world face increased competition to get grants. When grants are diverted for funding illusions, genuine research will be affected. By the time we understand the futility of those research findings, many laboratories around the world will have worked on the topic, resulting in global wastage. To cite an example, a few labs in Asia have already started research on persisters (Singh *et al.* 2009; Lou *et al.* 2008). One may argue that it indicates the importance of the topic and the validity of their findings. However, this is yet another instance of cajoling the scientific community through illusions. As indicated above, 28 years have passed without conclusively proving the ability of VBNCs to cause disease outbreaks, in spite of spending millions of dollars by a number of labs around the world. How much more time and money do we need to invest to prove a hypothesis?

Are we exaggerating our research findings to attract more attention?

In September 2009, the results of a $119 million HIV vaccine trial, carried out by the U.S. army and the Thai government, were made public (Butler 2009; Leavy 2009). The vaccine was given to 16,000 HIV negative people in Thailand who were at high risk of getting HIV. Half of the volunteers (8,197) received the vaccine, whereas the other half were given placebo. The participants were tested for HIV after three years. Results showed that 51 persons who received the vaccine got the infection compared with 74 patients who received the placebo. Thus, the study concluded that the chances of getting HIV were 31.2% less for those who had taken the vaccine. The results were hailed as a milestone in HIV vaccine research. However, not everyone was convinced. Some researchers have questioned the evidence from the data that the vaccine was protective while others have questioned the statistical methods used (Butler 2009). Dr. Adel Mahmoud (Princeton University, NJ) warns that raising expectations without scientific basis can be dangerous (Butler 2009).

Are the above results significant enough to be called a milestone? It would have been better had the researchers taken a cautious approach rather than claiming the findings as 'tantalizingly encouraging'. When will the scientific community admit that not all research can generate significant results?

Recently, a heated row over climate change made headlines following the leakage of e-mails after the computer server at the Climate Change Unit at the University of East Anglia was hacked. Skeptics claimed that the leaked e-mails showed data were being manipulated (http://news.bbc.co.uk/2/hi/uk_news/england/norfolk/8389727.stm). Similarly, the Intergovernmental Panel on Climate Change (IPCC) came under attack after a report from the IPCC published an exaggerated account on the melting of Himalayan glaciers (http://news.bbc.co.uk/2/hi/8468358.stm). The IPCC admitted the mistake later, but critics attacked the scientific credibility of the IPCC. Even though global warming can be real, there were suspicions that the data were manipulated or exaggerated for certain advantages. If mistrust in research findings exists, there is no one else to be blamed except the scientific community itself.

Why bubbles are created in scientific research? The most straightforward answer is that those who create bubbles are rewarded tremendously, both economically and academically. For example, the researchers who proposed VBNCs and persisters have been awarded millions of dollars in research grants. When the 'hotness' of the topic becomes the most important criterion for attracting grants, more topics will be 'made hot' either by manipulating data or by distorting the research findings.

In 2005, an article titled 'Why most published research findings are false?' appeared in PLoS Medicine (Ioannidis 2005). The author claims that most of the published research findings are false. Bias; testing by several independent teams; small studies and small effect sizes in a scientific field; greater numbers and less selection of tested relationships in a scientific field; greater flexibility in designs, definitions, outcomes and analytical modes; greater financial and other interests; and prejudices were all cited as the main factors leading to the publication of flawed research. It may be a matter of debate whether most of the findings are false; however it is certain that at least many of the findings are false. Unless certain practices in research are checked, a point will be reached soon when scientific research can provide only less for the money and the resources spent.

References

Butler, D. (2009). Jury still out on HIV vaccine results. *Nature* 461(7268), 1187-1187.

Chowdhury, M. A. R., Ravel, J., Hill, R. T., Huq, A. and Colwell, R. R. (1994). Physiology and molecular genetics of viable but non-culturable microorganisms. *In* "Biotechnology risk assessment: USEPA/USDA/Environment Canada: risk assessment methodologies" (M. Levin, Grim, C. and Scott, J. S., Ed.).

de Carvalho, L. B. (2006). Pressure also leads to worthless publications. *Nature* 439(7078), 784-784.

DeCoursey, T. E. (2006). It's difficult to publish contradictory findings. *Nature* 439(7078), 784-784.

Gannon, F. (2001). The essential role of peer review. *Embo Reports* 2(9), 743-743.

Ioannidis, J. P. A. (2005). Why most published research findings are false. PLoS Med 2(8: e124. doi:10.1371/journal.pmed.0020124

Lawrence, P. A. (2003). The politics of publication - Authors, reviewers and editors must act to protect the quality of research. *Nature* 422(6929), 259-261.

Leavy, O. (2009). Hiv Vaccine Results Controversy. *Nature Reviews Immunology* 9(11), 755.

Lou, C. B., Li, Z. Z., and Qi, O. Y. (2008). A molecular model for persister in E. coli. *Journal of Theoretical Biology* 255(2), 205-209.

Martinson, B. C., Anderson, M. S., and de Vries, R. (2005). Scientists behaving badly. *Nature* 435(7043), 737-738.

Oliver, J. D. (2005). The viable but nonculturable state in bacteria. *Journal of Microbiology* 43, 93-100.

Ravel, J., Knight, I. T., Monahan, C. E., Hill, R. T., and Colwell, R. R. (1995). Temperature-Induced Recovery of Vibrio-Cholerae from the Viable but Nonculturable State - Growth or Resuscitation. *Microbiology-Uk* 141, 377-383.

Richards, D. (2007). Little evidence to support the use of editorial peer review to ensure quality of published research. *Evid Based Dent* 8(3), 88-9.

Singh, R., Ray, P., Das, A., and Sharma, M. (2009). Role of persisters and small-colony variants in antibiotic resistance of planktonic and biofilm-associated Staphylococcus aureus: an *in vitro* study. *Journal of Medical Microbiology* 58(8), 1067-1073.

Wadman, M. (2005). One in three scientists confesses to having sinned. *Nature* 435(7043), 718-719.

Xu, H. S., Roberts, N., Singleton, F. L., Attwell, R. W., Grimes, D. J., and Colwell, R. R. (1982). Survival and Viability of Nonculturable Escherichia-Coli and Vibrio-Cholerae in the Estuarine and Marine-Environment. *Microbial Ecology* 8(4), 313-323.

INDEX

A

aging, 115, 117, 157–72, 179, 181,
    197, 198, 199, 201, 202, 206, 212,
    213, 214, 215
altered proteins, 226
aminoglycoside, 225
aminoglycosides, 17, 20, 23, 25, 27, 42,
    47, 57, 58, 75, 77, 81, 83, 84, 91,
    181, 185, 194, 195, 196, 197, 201,
    203, 205, 209, 210, 211, 212, 213,
    224
antibiotic pressure, 224
antibiotic tolerance, 41, 47, 49, 62,
    63, 179, 195, 201
antibiotics, X, 75–92, 170, 171, 179–
    211, 212, 213, 223, 224, 225, 226,
    230, 231
aquatic environment, 106, 119, 122,
    123, 124, 127, 130, 131, 133, 135,
    137, 224
asymmetrical, 103, 165, 166, 167,
    168, 169, 197, 225
ATP, 77, 78, 102, 160
AUC, 15, 209
AUC/MIC, 15, 17, 18, 19, 20, 59
autochthonous, 122, 130, 131, 132,
    133, 135, 137, 138, 224, 232
auxotrophic, 75, 76, 77, 78, 80, 81,
    88, 89, 92, 183, 196, 204, 205,
    207, 213, 215, 226
avirulent, 111, 212

B

bacterial heterogeneity, 179, 181, 201
bactericidal, 17, 22, 24, 39, 41, 47, 52,
    54, 58, 59, 61, 83, 90, 170, 191,
    213
bacteriostatic, 23, 24, 58, 59, 193
beta lactam. See lactam

beta lactamase. *See* lactamase
biofilms, 42, 43, 44, 45, 47, 50, 51,
    52, 54, 55, 59, 62, 64, 80, 81, 123,
    124, 180, 208, 209, 210, 230, 231
biological activity, 22, 23, 223

C

C. crescentus, 166, 167
carbonylation, 159, 165
CF, 44, 75, 80, 81, 82, 87, 89, 90, 207,
    208, 209
chironomid, 123, 130, 137
cholera, 113, 118–38, 231, 232
Cmax, 15
Cmax/MIC, 15, 17, 19, 20, 59, 209,
    210
coccoid, 102, 103, 106, 109, 117, 164
co-cultured, 86, 92, 209, 224, 231
concentration dependent, 17, 18, 55, 83,
    181, 185, 192, 209, 210
conditional senescence, 163, 164,
    165, 207, 225
copepods, 108, 120, 123, 130
crustaceans, 120, 122, 123
cystic fibrosis. *See* CF

D

D. melanogaster, 160, 166, 167, 168
daughter cells, 163, 165, 166, 167,
    168, 169, 170, 171, 172, 197, 198,
    206
debridement, 82, 88, 208
device related, 80, 81, 208, 209
dormant, 39, 40, 41, 48, 53, 54, 57,
    59, 63, 64, 91, 104, 109, 180, 223,
    225, 230

E

E.coli, 19, 20, 25, 27, 40, 43, 50, 51,
    76, 78, 101, 103, 107, 110, 111,
    112, 113, 116, 117, 125, 126, 137,
    161–72, 180, 181, 182, 184, 185,
    186, 187, 188, 189, 191, 194, 196,
    197, 203, 204, 206, 214, 225, 226
EF-Tu, 49, 63, 165
El Nino, 122, 136
electron deficient mutants, 85, 92,
    204, 205, 207, 210
electron transport, 77, 78, 79, 81,
    160, 196, 197, 212, 214, 215
ESBL, 21, 22, 24, 25
exponential phase, 22, 26, 39, 40, 41,
    48, 62, 125, 170, 180, 183, 187,
    188, 189, 192, 197, 204, 207, 213

F

F_oF_1-ATPase, 76, 78, 169
fatal infections, 82, 88, 91, 92, 224,
    231
fluoroquinolones, 17, 18, 19, 20, 23, 27,
    42, 44, 47, 54, 55, 209
FtsZ, 162

G

gentamicin, 22, 58, 77, 82, 88, 90, 91
gentamicin beads, 82, 88
GFP, 40, 181, 184, 199, 200, 201,
    211, 213, 214
glycogen granules, 124
gulls, 128

H

hemB, 77, 78, 82, 83, 84, 226
hemin, 43, 63, 75, 76, 77, 78, 79, 81,
    85, 88, 92, 196, 204, 215, 224,
    226
heritable phenotype, 179, 180, 193
hipA, 40, 41, 47, 48, 49, 50, 60, 62,
    63, 64, 180, 192, 223, 226
hipBA, 45, 49, 50, 63

hyperinfectious, 123, 124

I

inapparent, 131, 132
indigenous, 122, 124, 131, 135
injured cells, 105, 114, 115, 116, 117,
    127
inoculum effect, 21–25, 52, 56, 57,
    58, 62, 65, 223
inoculum size, 21, 22, 25, 26, 42, 52,
    53, 56, 57, 58, 63, 64, 91, 92, 182,
    187, 188, 190, 191, 192, 210, 211,
    223
inter-epidemic, 106, 121, 127

K

kanamycin, 51, 179–201, 203, 204,
    206, 210, 226

L

lactam, 20, 21, 23, 24, 26, 27, 52, 56,
    83
lactamase, 21, 22, 24
lifespan, 49, 50, 158, 159, 160, 161,
    163, 165, 166, 168, 169, 213

M

MazF, 46, 60
MazFE, 46
MBC, 15, 20, 21, 40, 41, 42, 51, 54,
    59, 203, 211, 226
MBC_{100}, 51, 59
membrane potential, 78, 79, 91, 160,
    196, 197, 201, 203, 205, 212, 213
menadione, 75, 76, 77, 78, 79, 81, 85,
    88, 89, 92, 215, 224, 226
menD, 77, 78, 83, 226
MIC, 15, 16, 17, 18, 20, 21, 22, 24,
    25, 26, 27, 40, 41, 42, 49, 51, 52,
    53, 56, 59, 64, 70, 192, 195, 196,
    199, 203, 209, 223, 226
mitochondria, 157, 158, 159, 160,
    161, 162, 168, 169, 199

mother cell, 165, 166, 167, 168, 169,
170, 171, 172, 197, 198, 206, 225
MPC, 18, 19, 20, 21, 223
MSW, 18, 19, 20, 21

N

noise in protein expression, 179, 181,
199, 200, 201, 211
non-dividing, 39, 42, 65

O

osteomyelitis, 44, 59, 75, 80, 81, 82,
88, 89, 207, 208
oxidative damage, 50, 116, 158, 159,
160, 163, 164, 165, 169, 213
oxidative phosphorylation, 159

P

PAE, 16, 25, 26, 27, 210, 211, 224
pandemic, 121, 122, 131, 132, 133,
134, 135, 136, 138, 224, 232
paradoxical effect, 52, 60, 61, 65, 192
PD, x, 15, 20, 25
penicillin, 19, 39, 47, 77, 79, 211
penicillinase, 21
persistent infections, 55, 56, 81
persisters, xi, 125, 170, 171, 172, 179,
180, 181, 182, 183, 184, 185, 186,
187, 188, 189, 190, 191, 192, 193,
194, 195, 196, 197, 198, 199, 200,
201, 202, 207, 210, 223, 225, 230,
231, 232
phenotypic shift, 39, 51, 52, 53, 54,
56, 57, 59, 60, 63, 64, 179, 180,
181, 190, 191, 192, 193, 201, 223,
230
PhoU, 50, 60
phytoplanktons, 119, 120, 122, 123,
137
PK, x, 15
PK/PD, 15, 16, 17, 19, 20, 21, 28, 56,
177, 207, 209, 223, 224, 226
planktonic, 42, 43, 44, 45, 51, 52, 57,
64

planktons, 120, 122, 129
post antibiotic effect. *See* PAE
pppGpp, 48
programmed cell death (PCD), 45,
46
protein oxidation, 116, 165, 168

R

rejuvenation, 167, 169, 170, 171, 197,
198, 206, 223, 225
relBE, 45, 47
replicative senescence, 163, 165, 166,
167, 169, 197, 202, 207, 225
resuscitation, 54, 56, 99–118, 119, 124,
125, 126, 127, 137, 205, 224, 230,
231, 232
resuscitation promoting factor (Rpf),
104, 105, 109

S

S. aureus, vii, 16, 18, 19, 20, 21, 22,
23, 24, 43, 52, 75, 76, 77, 79, 80,
81, 82, 83, 84, 87, 88, 89, 91, 104,
196, 214
S. cerevisiae, 165, 167, 168, 169
SCV, xi, 52, 75–92, 125, 177, 179,
195, 196, 197, 202, 203, 204, 205,
206, 207, 209, 210, 224, 226, 231
sediment, 108, 123
senescence, 157–72, 201, 202, 203,
205, 206, 211, 212, 214, 225, 226
senescent bacteria, xi, 65, 170, 172,
202, 203, 205, 206, 207, 210, 212,
213, 214, 215, 223, 225, 226
sequestrum, 208, 209
ship ballast, 123, 130, 131, 136, 137
slow dividing, 65, 91, 203, 204, 207,
210, 211, 214, 224, 225
small colony variants. *See* SCV
SOS, 48, 49, 60, 61, 63, 65
stationary phase, 22, 40, 41, 42, 43,
45, 46, 49, 50, 62, 63, 103, 114,
161, 163, 164, 165, 170, 180, 182,
183, 184, 185, 187, 188, 189, 201,
203, 204, 207

stochastic, 48, 157, 158, 199, 200,
    211
subpopulations, 18, 25, 39, 40, 41,
    42, 44, 46, 48, 51, 59, 60, 64, 75,
    105, 110, 111, 115, 116, 126, 170,
    179, 187, 188, 196, 202, 204, 210,
    211, 212, 214, 215, 224, 225
switching, 41, 77, 91, 157, 180, 207,
    224
symmetrical, 163, 166, 167, 169, 172,
    225

T

t>MIC, 15, 16, 24, 25, 28, 55
TA, 45, 47, 49, 50, 60, 192
TD-SCV, 87
temperature upshift, 107, 108, 111,
    112, 113, 114, 116, 117, 119, 126,
    127, 137
thiamine, 75, 76, 77, 92, 181, 183,
    196, 197, 224
thyA, 77, 78, 87
thymidine, 75, 76, 77, 85, 89, 92, 113,
    181, 183, 196, 197, 215, 224
time dependent, 17, 55, 117, 191
T_{MSW}, 19, 20
toxin-antitoxin. *See* TA

Type-I persisters, 40, 41, 180, 207,
    223
Type-II persisters, 41, 180, 207, 223

U

ultrastructure, 79, 206
unsaturated fatty acids, 76, 78

V

V. cholerae, 101, 103, 106, 108, 113,
    117, 119–38, 224, 232
V. parahaemolyticus, 108, 114, 115,
    119, 122, 128, 136, 137
V. shiloi, 129
V. vulnificus, 101, 102, 104, 105, 107,
    108, 109, 111, 114, 115, 117, 119,
    127, 128
VBNC, 99–138, 179, 189, 201, 202,
    205, 206, 207, 224, 231, 232
viable but non culturable bacteria.
    See VBNC

W

water birds, 123
winter, 101, 106, 111, 119, 121, 127,
    128, 129, 130, 131

www.ingramcontent.com/pod-product-compliance
Lightning Source LLC
Chambersburg PA
CBHW071412170526
45165CB00001B/251